ARAB MILITARY INDUSTRY

Capability, Performance and Impact

Also available from Brassey's

BELLAMY
Expert Witness: A Defence Correspondent's Gulf War

COOLEY
Payback: America's Long War in the Middle East

CORDESMAN
Weapons of Mass Destruction in the Middle East

EHTESHAMI
Nuclearisation in the Middle East

ISRAELI
Muslim Fundamentalism in Israel

KARSH & RAUTSI
Saddam Hussein: A Political Biography

McMICHAEL
The Stumbling Bear: Soviet Military Performance in Afghanistan

ORMAN
Faith in G.O.D.S.: Stability in the Nuclear Age

Centre for Arab Unity Studies

ARAB MILITARY INDUSTRY

Capability, Performance and Impact

YEZID SAYIGH

Published in association with the Centre for Arab Unity Studies
(Beirut, Cairo and London)

BRASSEY'S (UK)

LONDON · WASHINGTON · NEW YORK

First English edition 1992

UK editorial offices: Brassey's, 165 Great Dover Street, London SE1 4YA
orders: Marston Book Services, PO Box 87, Oxford OX2 0DT

USA editorial offices: Brassey's, 8000 Westpark Drive, First Floor, McLean, VA 22102
orders: Macmillan Publishing Company, Front and Brown Streets, Riverside, NJ 08075

Distributed in North America to booksellers and wholesalers
by the Macmillan Publishing Company, NY 10022

Library of Congress Cataloging in Publication Data
available

British Library Cataloguing in Publication Data
A catalogue record for this book is
available from the British Library

ISBN 0-08-041777-9

Typeset by Florencetype Ltd, Kewstoke, Avon
Printed in Great Britain by BPCC Wheatons Ltd., Exeter

In Memoriam

Riad Al-Ashkar,
b. Jerusalem 1947, d. Beirut 1982

Preface

The first draft of this book was prepared some months before the Iraqi invasion of Kuwait in August 1990, but the text has since been revised and updated to take account of the Gulf crisis and war, and of subsequent events up to October 1991.

Originally written for the Centre for Arab Unity Studies (Beirut), the Arabic version of this book contains additional chapters that deal with Israel and other comparative cases of Third World arms-producers – India, Brazil, Iran, Turkey and Pakistan – and that propose a strategy for Arab military industrialization. I have excluded these chapters from the English version, mainly for reasons of space but also because there is already a substantial body of writing on most of the other countries mentioned, making the comparison less useful for an English-reading audience.

Various people provided me with valuable research assistance while I was preparing this book. In particular I wish to thank the staff at the libraries of the Centre for Arab Unity Studies in Beirut, al-Ahram Strategic Studies Centre in Cairo, and the International Institute for Strategic Studies in London, among others. Haifa Sayegh and her colleagues at the library of the League of Arab States in Tunis cheerfully responded to my numerous requests for material and made me feel at home during my repeated visits.

I also owe a special debt of gratitude to Heino Kopietz for sharing with me his own archival material and his knowledge of Middle East defence budget accounting. My father, Yusif Sayigh helped me understand the more complex industrial and economic concepts, while Tony Zahlan and Nabil Sha'th willingly discussed the problems and pitfalls of science and information in the Arab World. I am further grateful to former Egyptian Minister for Defence and Defence Production, Amin Hewedy, for giving me the benefit of his experience, as I am to the participants in the special panel on Arab Military Industrialisation convened by the Centre for Arab Unity Studies, in Cairo, in November 1988. Finally, special thanks go to Khaireddin Hasib, director of the Centre, both for asking me to write this book and for his patience and understanding as deadlines came and went.

As ever, Liz, Serine and Yusif surrounded me with love and support during the preparation of this book. Yusif was born just as I started writing, reminding me once more that there are far more important things in life than arms industries. For their sake and mine I hope the message gets through.

Yezid Sayigh
October 1991

Contents

List of Tables

List of Abbreviations

a/c	Aircraft
AA	Anti-Aircraft
AAM	Air-to-Air Missile
ACC	Arab Co-operation Council
AD	Air Defence
AFV	Armoured Fighting Vehicle
AIC	Advanced Industrialised Country
AIPAC	American Israeli Public Affairs Committee
ALESCO	Arab League Education, Science and Culture Organisation
AMU	Arab Maghreb Union
AOI	Arab Organisation for Industrialisation
AP	Anti-Personnel
AP	Armour-Piercing
APC	Armoured Personnel Carrier
AT	Anti-Tank
ATGM	Anti-Tank Guided Missile
AWACS	Airborne Warning and Control System
BAe	British Aerospace
C^3I	Command, Control, Communications and Intelligence
CEP	Circular Error Probablility
CSCE	Conference on Security and Co-operation in Europe
DIDA	Defence Industry Support and Development Agency (Turkey)
DIPF	Defence Industry Promotion Fund (Turkey)
ECM	Electronic Counter-Measures
EW	Electronic Warfare
FAC/M	Fast Attack Craft, armed with Missiles
FAE	Fuel-Air Explosives
FCS	Fire Control System
FMS	Foreign Military Sales (USA)
FTZ	Free-Trade Zone
FY	Fiscal Year
GCC	Gulf Co-operation Council
GDP	Gross Domestic Product

GDR	German Democratic Republic
GEMI	General Establishment for Military Industries (Saudi Arabia)
GIAT	Groupement Industrielle d'Armements Terrestres
GNP	Gross National Product
HE	High Explosive
HEAT	High Explosive Anti-Tank
HUD	Head-Up Display
IAF	Israeli Air Force
IFF	Identification Friend or Foe
IFV	Infantry Fighting Vehicle
IR	Infra-Red
ISI	Import-Substituting Industrialisation
JCSS	Jaffa Center for Strategic Studies
JTG	Jordanian Technology Group
LDC	Less Developed Country
MAP	Military Assistance Program (USA)
MBT	Main Battle Tank
MIA	Military Industrialisation Authority (Iraq)
MICV	Mechanised Infantry Combat Vehicle
MoU	Memorandum of Understanding
MRL(S)	Multiple Rocket Launcher (System)
NBC	Nuclear, Biological, Chemical
NIC	Newly-Industrialising Country
NODP	National Organisation for Defence Production (Egypt)
OAPEC	Organisation of Arab Petroleum-Exporting Countries
PGM	Precision-Guided Munitions
PLO	Palestine Liberation Organisation
RAP	Rocket-Assisted Projectile
ROF	Royal Ordnance Factories
RPV	Remotely Piloted Vehicle
SA	Sud Aviation
SAM	Surface-to-Air Missile
SCA	Supreme Committee for Armament (Egypt)
SDI	Strategic Defence Initiative
SIDF	Saudi Industrial Development Fund
SIPRI	Stockholm International Peace Research Institute
SP	Self-Propelled
SSM	Surface-to-Surface Missile
STOL	Short Take-Off and Landing
UAE	United Arab Emirates
UNESCO	United Nations Education, Science and Culture Organisation
V/UHF	Very/Ultra High Frequency
VTOL	Vertical Take-Off and Landing

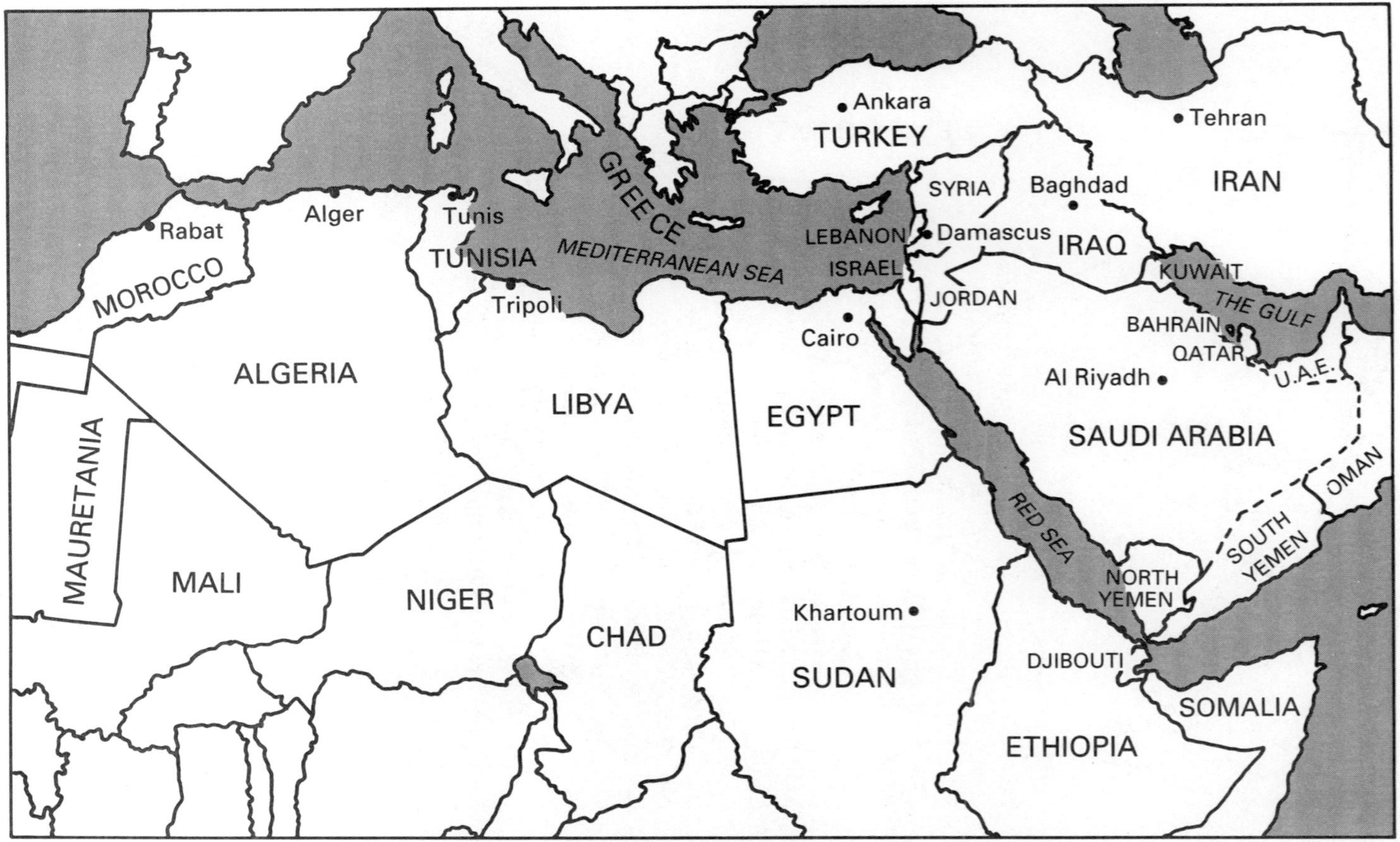

The Middle East

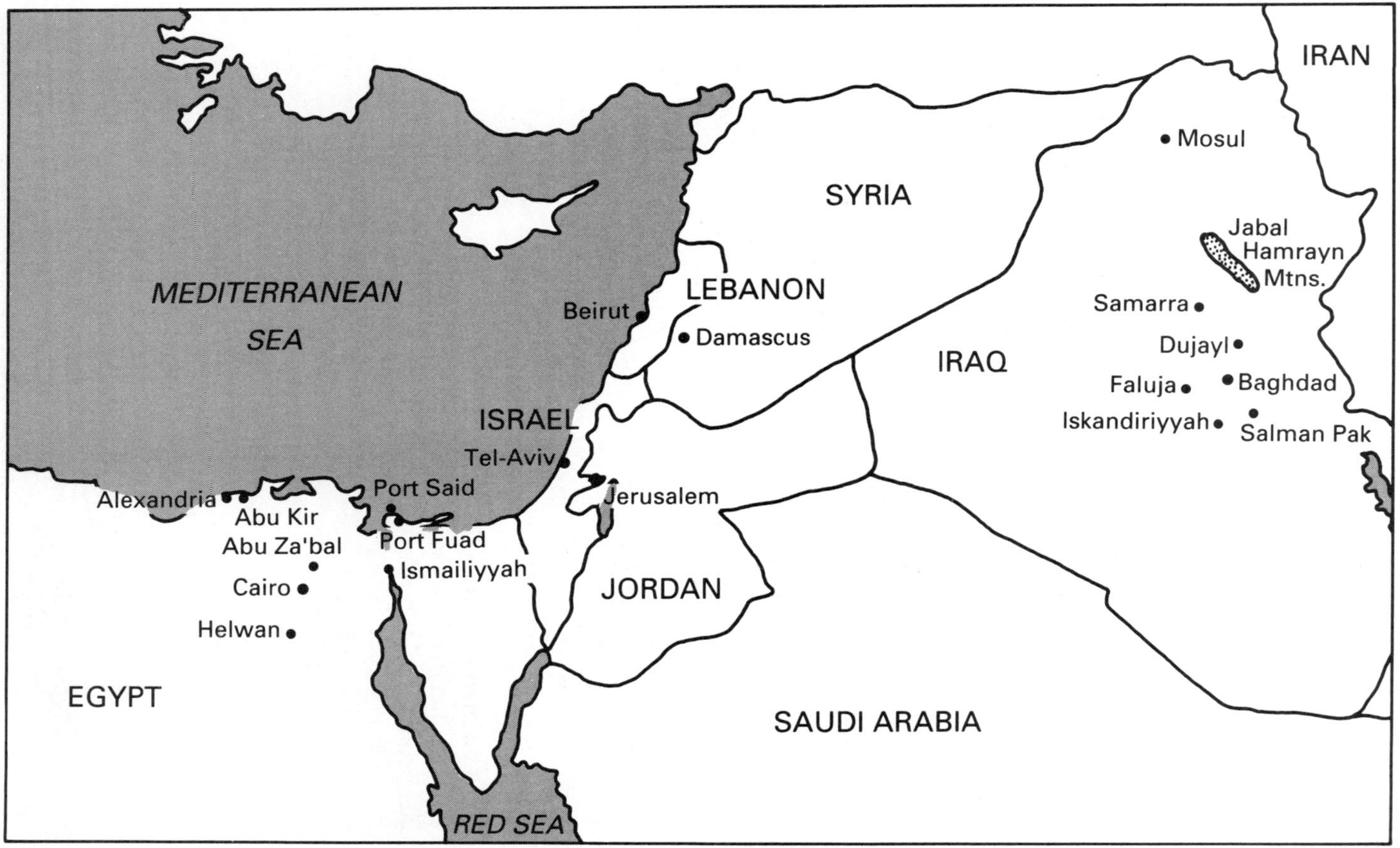

The Arab East (Mashreq)

1

Introduction: Incentives and Determinants of Third World Defence Production

A growing phenomenon in international security is the increased activity of Third World arms producers and exporters. From a mere handful in the 1950s, the number of developing countries involved in major defence production reached around 20 in the late 1980s, with as many again undertaking limited activities, such as the manufacture of small arms and light ammunition. Although at 1.5–3 per cent during the 1980s their share of total world military trade remained minor, its importance for their economies and its impact on local conflicts has been more significant.

The Arab countries have witnessed a similar development, with Egypt long-established as an important arms-producer and Iraq (at least until the Gulf war of 1991) and Saudi Arabia proceeding with their own plans for major military industrialisation. In addition, half a dozen other Arab states have undertaken limited, relatively small-scale production at one time or another in the past two decades. This has taken place against a background of extremely high defence spending, accompanied by a high level of arms imports that has made the Middle East region the largest single recipient of arms transfers within the Third World. The overwhelming importance of ensuring defence and reinforcing national security, coupled with the costs of weapons acquisitions, has encouraged Arab military industrialisation as a means of reducing political and strategic vulnerability to foreign suppliers and of improving the economics of military expenditure.

Despite these sweeping opening statements, little systematic study has actually been made of the incentives, performance, and prospects of Arab defence production.[1] It is the purpose of this book to redress this situation and present a complete picture of the present reality of Arab defence industries, as well as an assessment of their technical, operational, and systemic dimensions. This also involves an evaluation of their economic impact, and an examination of the requirements of wider and more effective military industrialisation. It cannot be emphasised enough, however, that relatively little information has been made public by the Arab governments and industrial organisations concerned regarding output, costs, labour, technological and manufacturing levels, and research and development (R&D). What is available is often undependable or inconsistent, and the researcher is compelled to draw on a variety of sources, including newspaper and magazine articles, although they may not be equally reliable. Great effort has been exerted to corroborate information and ensure its credibility and consistency, but the reader should keep this note of caution in mind.

This book is divided into four main parts. Part One deals with the theoretical and strategic context of Arab arms production. The classic arguments concerning incentives and disincentives of Third World military industrialisation are discussed in the introduction to offer an overall framework for studying the Arab case. Chapter 2 provides the strategic context by reviewing the patterns of defence spending and arms imports in the Arab region. A critical assessment of the trends in Arab military development follows in Chapter 3, in order to identify areas of need and set the basis for evaluating indigenous industrial efforts.

Part Two examines local defence production in the Arab countries. It opens with Chapter 4, which provides a country-review of Egypt, the largest Arab arms producer, followed by Iraq and Saudi Arabia in Chapters 5 and 6. Chapter 7 summarises industrial developments among the remaining, smaller Arab arms producers.

Part Three assesses Arab military industrialisation on the basis of the preceding chapters. Within it, Chapter 8 focuses on technical and operational dimensions, in which the role is evaluated of inputs such as technology transfer, civilian industry, manpower, and exports. Chapter 9 then discusses the balance between security and economic incentives as the main driving force for Arab military industrialisation, examining linkages between the defence sector and the civilian economy as a whole. Finally, Chapter 10 reviews the Arab scientific and industrial base, which ultimately forms the key determinant in the long-term success or failure of indigenous defence production.

Part Four pulls the patterns together and draws out their implications. Chapter 11 re-examines the theoretical issues and interrelationships involved, as well as the security and economic incentives that will continue to drive Arab military industrialisation in the future. The argument is refined in Chapter 12, which debates the assumptions, political framework, and possible aims and needs of Arab defence production and then defines the determinants governing effective industrial performance. An Epilogue discusses the implications and prospects for Arab military industrialisation in light of the Gulf crisis and war of 1990–91.

General Incentives for Military Industrialisation

A variety of incentives have been suggested as main reasons for the trend towards indigenous defence production in the Third World. These group easily within two broad categories: national security and economic benefits. The literature on the subject is varied, but certain elements are now standard and may be summarised as follows:

National Security and Defence

A primary motivation for the development of local arms industries in Third World countries is the wish to reinforce political independence. The assumption is that self-sufficiency in providing their own needs of military hardware reduces their military and strategic dependence on outside powers.[2] The USA, the USSR, and other advanced industrialised countries (AICs) exert considerable influence over recipients among the developing countries, especially when the latter rely on a single main source of arms.

Perceptions of external threat, which are a major driving force behind most regional arms races in the Third World, are an additional incentive for local manufacture of weapons.[3] This is particularly the case if external suppliers are likely to cut off shipments in the run-up to, or during, inter-state conflict. A third incentive is the desire of governments to enhance regional standing and power projection capability; to this should be added their parallel desire to increase political prestige domestically.[4]

Arab analysts and officials have further refined and developed these now-standard tenets. In forming the Arab Organisation for Industrialisation (AIO) in 1975, its four member states stated a number of general aims, foremost of which were: to provide a measure of self-sufficiency in conventional hardware and so enable Arab armies to fight longer without foreign resupply; and to contain Israeli or other external, political and strategic pressures.[5] In the 1980s, the rapid expansion of the Iraqi defence industry was fuelled by the concern to achieve the following objectives: circumvent control or interference from supplier countries; avoid vulnerability to arms embargoes; pre-empt political pressure on suppliers from third parties (such as Israel or Iran); and ensure the secrecy and security of local defence needs and projects.[6] The last is a particularly relevant consideration, given the sensitivity of security and defence in an Arab environment in which there is heavy reliance on foreign experts, trainers, advisers, and technicians in the operation, maintenance, and support of hardware, as well as in the provision of other services (such as construction).

Several analysts note additional incentives and factors influencing the decision to initiate or expand local defence production. At the military level, a local R&D or production capability allows users to adapt hardware to local combat needs and field conditions, as well as directing industrial activity according to domestic economic and social circumstances and priorities.[7] At the political level, several groups of people – such as senior army officers, private industrialists, and public sector managers – may have a vested interest in the initiation or continuation of indigenous defence production, which they influence by wielding considerable decision-making power.[8]

Economic Arguments

The economic benefits of local arms production are equally stressed by Third World officials and experts, often as the major motivation. This may initially be a method for obtaining approval and funds, though once activity has started or economic pressures mount, economic and commercial considerations gain added importance, if only as a further argument to ensure continued government commitment to the substantial investment required in the sector. One study asserts that the most important economic incentive for defence production is that it should substitute for imports.[9] This allows hard currency savings and thus an improved balance of payments – especially important at a time of rapidly spiralling costs of imports and high-tech products – while the trade balance is further improved if exports can be achieved at a later stage.

A second claimed economic benefit is to utilise any excess of capacity in the civilian industry and so increase the dynamic linkages between the defence and the civilian sector. These are especially relevant, thirdly, in the case of generating spin-offs from arms production for the rest of the economy. These include induced R&D, increased labour productivity, and acquisition of skills and technology familiarisation.[10] Finally, governments can work to achieve economic stabilisation by directing military industrial contracts and investment, reducing imports or overall defence spending, or stimulating growth through new weapons production.[11]

In taking up these economic incentives, some Arab analysts have offered additional dimensions. In the case of the AOI, for example, emphasis was also placed on the use of defence production as a basis for advanced industrialisation in the Arab region, which would offer an opportunity to acquire technology and train manpower. The Iraqi analysis cited above offered further refinements, in addition to the above aims: absorb and utilise surplus labour; activate associated industries, such as steel and chemicals production; increase the share of locally-added value and so improve GDP; utilise national resources – human, financial, and natural – better and so reduce dependence on the outside and vulnerability to world fluctuations.[12]

Counter-arguments

The arguments in favour of Third World military industrialisation are by no means unchallenged. One of the most forceful objections is that the high technological content of modern weapon systems leads to cyclical obsolescence, which substantially raises the cost of local R&D and production. However, financial cost is not the real problem, but the fact that the need to keep pace only increases dependence on foreign suppliers. Indeed, *technological* dependence may lead to a position of far greater subservience to external sources than arms dependence or even than the old colonial relationships, in the view of some observers. This is because the former is built into the entire productive system at all levels yet is inherently impossible to master locally in full.

This is the thrust of counter-arguments by Vayrynen and Kemp among others.[13] Among Arab analysts, Amin Huweidi stresses this aspect in order to conclude that local arms production is essentially counterproductive and should be abandoned in favour of diversification of sources of arms imports.[14] He notes that 'the mirage of indigenous arms production . . . usually ends in the most tyrannical kind of dependence named military technical dependence'.[15] This may not apply to some of the simpler 'combat consumables' – such as infantry and medium weapons and their ammunition, and other types of ordnance from mines to air bombs – which require low technology, but even for these the ability of the local industry to attain surge-production during sudden crises can be limited.[16] In any event, while displaying concern about the eventual deepening of technological dependence and its consequences, Western observers are not as adamantly opposed to indigenous defence production as Huweidi is, with some analysts noting that diversification of arms sources carries its own price in the form of greater expense in stockpiling parts and the provision of follow-on support services.[17]

Despite the validity of these objections and the extreme seriousness of technological dependence, most negative commentary focuses on the adverse economic consequences of Third World military industrialisation, or more specifically on the general paucity of the benefits promised. Most prominent is the distinct absence of tangible spin-offs in the form of R&D and training for the civilian sector, even in

Third World countries that have invested heavily in R&D and industrial development (such as India) or in the newly-industrialising countries (NICs), such as Brazil or South Korea.[18] Deger also notes that more advanced and productive methods of arms production tend to be capital-intensive, which is particularly unsuited for labour-intensive countries like India and Egypt.

A contradiction arises in this context: a main drawback of arms imports is that they form a significant share of intermediate-level capital goods, and thus effectively reduce industrial output and productivity; yet the local manufacture of arms does not necessarily remedy the situation as it too relies heavily on imports of production machinery and tooling, as well as on primaries and components.[19] Indeed, a telling technical and financial criticism of local defence production is that it may actually raise overall imports and hard currency expenditure.

Determinants

The economic counter-arguments extend further. The main additional criticism is, in fact, connected to a basic determinant of successful defence production in any country: the civilian industrial base. In the words of Brzoska and Ohlson, 'Arms production on a large scale requires the existence of a sufficiently large and differentiated industrial base'.[20] It is this element that allows dynamic linkages to be set up between the civilian and the military sector, but it is also one of the most under-developed areas of economic activity in the developing countries. Thus the linkage may be positive, but the mismanagement that often occurs can actually lead to increased costs.[21] By implication, moreover, the progress of military industrialisation is very much a function of a given country's overall industrial strategy, and specifically of its approach towards import-substitution or export-generation.[22]

In addition to the potential capacity of the civilian industrial sector to sustain a military-industrial complex, Deger identifies two additional, main determinants of Third World arms production. These are the opportunity costs of the resources involved and the various types of spin-off that may occur.[23] The latter may be debatable, though, as a number of Third World arms producers have demonstrably been able to pursue military industrialisation despite minimal spin-offs. More pertinent might be Wulf's characterisation. He defines three central questions that determine the extent to which arms production in a Third World country is possible: (1) how diversified is the country's industry, and does it have a minimum level of R&D and skilled manpower; (2) can production costs be kept low, and to what extent will local resources be allocated to the sector; and (3) does the country have an indigenous development and production capacity, and how can it secure design and process technology?[24]

From that starting point, Wulf goes on to define two key indicators of a country's ability to produce arms.[25] One is the output of six central industries: iron and steel; non-ferrous metals; metal products; non-electrical machinery; electrical machinery; and transportation equipment. The second indicator consists of two types of manpower: the number and skills of the persons employed in the central industries, and the number of scientists, engineers, and technicians involved in R&D. These indicators evidently affect productivity, as well as the ability of the civilian industry to sustain defence production. They also affect the ability of that country to produce at least a part of its own manufacturing machinery and tooling, an item that is often more difficult to obtain and as costly as system designs or actual hardware. The aspect of R&D, moreover, determines what level of technology transfer can be achieved, as it sets local ability to absorb and adapt foreign inputs.

Taken together, it is the combination of the determinants suggested in the preceding paragraphs that places a given country on the ladder of military industrialisation. This comprises seven main rungs: (1) servicing and repair of imported weapon systems; (2) overhaul of imported weapon systems; (3) local assembly of imported subassemblies; (4) limited licensed production with assembly of some locally-made components, and possible re-export to licenser; (5) independent licensed production but with continued import of components; (6) local R&D leading to modification of licence-produced arms; and (7) complete independence in R&D and production.[26] Of course, the local industry may operate at several different stages at once, with different programmes occupying several rungs on the ladder. However, it is virtually impossible for any developing country to reach the seventh stage, as it is only the two superpowers that can still afford to produce wholly indigenous weapon systems.

Summary

The preceding discussion covers the main issues – incentives and determinants – that shape the process of military industrialisation in developing countries. This book examines the continuing debate about the pros and cons of indigenous defence production in the Arab case, in order ultimately to make three core arguments.

The first is that in the competition between national security considerations and economic incentives, it is the former that have provided (and will continue to provide) the main driving force for Arab military industrialisation, whereas the latter remain of secondary importance. Secondly, that the Arab arms producers enjoy medium-to-low ability to undertake successful arms production, measured by the criteria described above, although there is scope for major savings and spin-offs to be achieved by focusing on specific areas of technology, hardware, and support services rather than the more prestigious high-tech, front-line systems. The third argument is that, far from actively promoting development, Arab defence industries are not only seriously handicapped by the under-development of their civilian economies but are unlikely to develop fully unless broader economic, social, and even political development is attained. A corollary is that a restructuring of local defence production capabilities is possible, which will both contribute appreciably to GDP and assist development by providing certain categories of hardware and services.

PART ONE

The Strategic Context

2

Arab Arms Imports: Strategic Context and Patterns

Introduction

There can be little dispute that the Middle East as a whole has been the largest single recipient of world arms transfers within the Third World. According to SIPRI, which takes the region to include Israel and Iran, it accounted for 44.4 per cent of all major weapons sales to the developing countries in the period 1969–88.[1] This trade was worth $166.82bn (out of a total of $375.92bn) at constant 1988 prices. However, these figures reveal less than the whole picture, because SIPRI excludes the North African and the sub-Saharan (Horn of Africa) Arab states from its definition of the Middle East. Adding the former group alone brings the total value of that region's arms imports in 1969–88 up to $200.85bn, or 53.4 per cent of the developing countries' total for the two decades.

A detailed breakdown is not available for the 1969–88 period, but in 1971–85 the Arab states alone (excluding Iran and Israel) accounted for 44.5 per cent of total arms imports to the Third World.[2] The value of Arab purchases was $127.26bn. Applying the same proportion to the longer 1969–88 period points to imports worth $165.4bn. A narrowing down of the focus shows that of the top 25 Third World importers in 1976–85 there were 10 Arab states, which received 44.05 per cent of all arms transfers from all sources to that group (estimated at a value of $94.93bn, at constant 1985 prices).

The Strategic Context

The volume of this trade in arms reflects the state of conflict and the insecurity in the region, that have led both to war and to regional arms races. As important, most countries in the Middle East region were able, until the early 1980s, to finance their purchases thanks either to oil wealth or to subsidies from external allies. The decline in oil prices subsequently reduced the volume of the flow of weapons to the area, with effect not only on the producers but on their neighbours as well. It should be noted, however, that total figures for Arab arms imports remained high after 1980–82, against expectation, owing largely to Iraqi expenditures during the Iran-Iraq War and the Syrian build-up following the Israeli invasion of Lebanon.

The Gulf crisis and war of 1990–91 have raised inescapable questions about the future flow of military technology to the Middle East and the trend in oil prices and revenues. Yet even before 1990, a revival of the arms trade to the region was not inevitable, despite prospects of a rise in oil prices. This suggests that there are other, systemic factors that are tending to limit further arms transfers.

Among the reasons for the slow-down of purchases by most Arab countries in the 1980s was the saturation of their armies with modern weapon systems. The latter require time to be fully absorbed and utilised, and it may be another few years before Arab armies are ready to undertake another major wave of acquisition and absorption. Indeed, as arms become increasingly sophisticated technologically, the time-lag may even increase, with far more time passing between the introduction of successive generations of systems. The result may be a pattern of arms imports that is much reduced in overall volume and which is, moreover, spread out over longer periods of time. The immediate reactions to the Gulf war – massive new Saudi requests for American weapon systems and the approval of new American sales to Egypt (and Israel) – may not indicate a radical and sustained upsurge in imports in the near future, after all.

Two other issues will also affect the pattern of arms sales and imports in the Arab region in coming years. For one, the dynamics of conflict and crisis management in the Middle East are changing and may eventually defuse the regional arms race, especially if the opportunity paradoxically offered by the Gulf crisis is exploited. The proposals suggested by President Bush at the end of May 1991, for regional arms controls in both conventional and non-conventional weapons spheres, may herald a new phase of international cooperation by the major producers and suppliers of advanced military technology to control local arms races and their destabilizing effects. The second issue that may affect Arab arms patterns is the potential impact of a developing, indigenous, defence production capability.

The Iraqi invasion of Kuwait and the inability of the Arab states to resolve the issue (peacefully and without intervention) cast major doubts about the efficacy of inter-Arab regional co-operation and the future of existing collective bodies. The Arab Co-operation Council (ACC), founded as recently as February 1989, was effectively dissolved, while the League of Arab States split down the middle. Even the Gulf Co-operation Council (GCC) and the Arab Maghreb Union (AMU), both of which survived intact, failed to act as policy-coordinating or decision-making agencies. The immediate prospects for collective security based on local states are not encouraging – although the GCC member states, along with Egypt and Syria, established a shortlived 'nucleus of an Arab defence framework' in March 1991[3] – yet bilateral pacts with Western or other external powers may prove of limited use as a support or added deterrent.[4]

Nonetheless, Arab collective security may yet re-remerge, in part under the impetus of an Egypt anxious to assert its regional influence by addressing the issues of Palestine and Gulf security and revitalising the League of Arab States.[5] Moreover, the fact that the Arab countries are faced with common social, political, and economic problems may prove to be an incentive for regulating inter-state conflict and curbing the regional arms race, partly in order to reduce defence spending.

The Gulf crisis and war have additionally contributed to a growing international climate that is conducive to arms control. The tendency of the USA and the USSR to agree on regional issues – dramatically revealed during the Gulf crisis – constrains the margin of freedom for Third World actors to pursue their local conflicts or to involve their superpower patrons. The ability of the USA to mobilise the leading Western nations and utilise the United Nations framework in support of its policy signals a further constraint. It is not inevitable that either superpower will reduce arms transfers, though, especially with the presence of so many other suppliers ready to fill the vacuum. However, external agreement may limit the flow of certain weapon types or reduce resupply during conflict, thus discouraging local state actors from initiating hostilities at all.

Furthermore, in referring to the regional strategic environment and its effect on the arms trade, the impact of the proliferation of weapons of mass destruction and the emergence of a regional balance of terror in the Middle East has yet to be seen. The Gulf war has stimulated intense efforts, particularly in the West, both to dismantle Iraq's non-conventional capability and to constrict the flow of nuclear, biological and chemical (NBC) and missile technology or materials to the region. Whether or not this will extend to include other states with NBC or ballistic missile capabilities (such as Israel) remains to be seen; nor are the implications of maintaining the existing stalemate (with Israeli nuclear dominance) clear.

Either way, the perpetuation of the post-Gulf-war, non-conventional balance – with disarmament of Iraq but retention of Israeli capability – or renewed proliferation will not necessarily lead themselves to a decline in the need for conventional weapons and large standing armies. The assumption here is that a credible defensive force must be retained in order to deter or defeat attacks that, though serious, are not threatening enough to warrant a resort either to nuclear/chemical weapons or to ballistic missiles aimed at population centres. Indeed, it is conceivable that the existence of an 'ultimate' deterrent might even encourage the start of war, as it is seen to impose a 'ceiling' on operations. By artificially 'freezing' the balance of power at a certain level, non-conventional deterrence allows the weaker party to wage the battle on more even terms. The Gulf war has not obviated either perception: conventional force decided the conflict, and Israeli nuclear strength did not prevent Iraqi missile attacks (though it may well have deterred chemical attacks).

The second issue affecting future Arab arms imports is the growth of local productive capabilities. Here again it is difficult to predict, but some observations may be made. Although major weapon systems are the most prominent component of imports in terms of visibility and unit cost, it is not these that form the bulk of such procurement. Rather, the largest part of Third World arms trade consists of medium and small arms and ammunition, personal gear, combat support hardware, electronic systems, communications equipment, spare parts, maintenance, training, logistics and support, and construction or engineering/design services.

The clear implication here is that major savings can be, and in certain cases have been, made in the latter categories by Arab or other Third World states that seek to develop local substitutes. This is not to deny that added savings could conceivably be made by producing certain major weapon systems locally too, since that would reduce imports and preserve hard currency. But the suggestion is that, even if technical difficulties inhibit the production of such high-tech weapons that involve complex components and sophisticated manufacturing processes, indigenous arms production can still have a beneficial impact on imports and currency earnings. This is especially true in periods when the Arab countries are still absorbing new major equipment and so do not need to buy additional systems for some years, but when they nonetheless need continued supplies of small arms, spares, services and the like.

Arab Arms Imports

The foregoing comments suggest the need to review the main patterns in Arab arms imports, as they directly affect the nature and orientation of local military industry. The following sections analyse the arms trade according to its distribution by type (of hardware), recipient, and supplier.

Distribution According to Type

Any discussion of Arab arms imports runs into several difficulties. One problem is that public scrutiny of the trade tends to focus on major weapon systems. Relatively 'minor' items of combat hardware or non-combat equipment and military technology or components are not as regularly reported, if at all. In these categories, commercial deals may pass unnoticed or be channelled through several agents and intermediaries between the original supplier and final customer – unlike major weapons which are difficult to conceal. Another problem is that many countries – both suppliers and recipients – offer no official details on numbers or prices involved, making it necessary to rely on estimated values. Further grey areas exist in deciding what to consider as imports and what not to when it comes to items that are partly produced locally, under licence. Nonetheless, a review of the pattern of major weapons imports into the Arab region gives an idea of the scale of the military trade and therefore of the technical and financial issues involved in seeking either reduction or substitution.

The year 1973 is a natural starting point for a comparative look at contemporary Arab military development. The fourth Arab–Israeli war of October and the revolution in oil prices that followed combined to fuel a massive build-up in most countries of the region. Table 2.1 summarizes the main holdings of major weapon systems in all Arab countries in two years for contrast: 1973, just before the rearmament and spending process, and 1990.[6] Among principal land weapons, the number of tanks rose by 14,062 (205 per cent), other combat armoured vehicles by 25,999 (365 per cent), and artillery pieces by 9,045 (392 per cent). Over the same 16-year period, the number of combat aircraft rose by 1,535 (85 per cent), transport aircraft by 196 (82 per cent), trainers by 991 (278 per cent), and helicopters by 1,394 (249 per cent). At sea, the number of major surface combatants (frigates, destroyers, and corvettes) increased by 34 (262 per cent), submarines by 13 (108 per cent), and attack and patrol craft (generally those above 80–100 tons) by 130 (62 per cent).

Massive as the above figures may be, the real numbers for Arab weapons acquisitions in the period 1973–90 are significantly higher in fact. The inventories cited in Table 2.1 refer to holdings at specific moments: 1973 and 1990. However, many systems in 1973 later became obsolete and were gradually removed from service in the 1970s and 1980s. These had to be replaced. Similarly, replacements were acquired to make up for the heavy losses incurred during the 1973 and the 1982 war. Thus the total figure for, say, tanks in 1990 reflects the *net* increase over 1973: it does not reveal the additional number of tanks that were lost in the interval through combat, cannibalisation, or obsolescence and phasing out. True, figures for 1973 were often vague or imprecise, and so part of the apparent increase in inventories in 1990 may simply have been due to better reporting. And since it is not always possible to tell which weapons were lost in combat and which simply removed from service, there is the risk of double counting. To allow for these problems, a conscious effort has been made to err on the side of caution, by underestimating replacement acquisitions.

Nonetheless, reasonably accurate, albeit somewhat tentative, totals can be arrived at for arms acquisitions by the Arabs since 1973. Calculations based on known combat losses and on identification of specific weapon systems known to be in or out of service at given times suggest that more realistic figures for the overall increase in principal combat weapons acquired between 1973 and 1990 are as follows: 15,962 tanks (232 per cent), 27,556 other combat armoured vehicles (387 per cent), 9,563 artillery pieces (414 per cent), 2,423 combat aircraft (134 per cent), and 206 attack and patrol vessels (99 per cent). These figures are compared in Table 2.1.

It should be stressed, again, that the figures in the Table 2.1 refer only to selected major weapons systems, and do not include many other types. Excluded are support and transport vehicles, multiple rocket-launchers and mortars (which are an integral and important part of artillery forces), anti-tank guns and guided missiles, anti-aircraft guns and guided missiles, electronic-warfare aircraft, light coastguard craft, and auxiliary naval vessels (such as training, support, mine warfare, and amphibious/landing craft), among others. The cost of acquiring such equipment is, of course, substantial.

Comparison of the holdings cited in Table 2.1 with those of principal member countries of NATO and the Warsaw Pact (excluding the two superpowers) places Arab inventories in even clearer perspective. No major conclusions can be drawn from the contrast in Table 2.2, but it serves to stress the sheer scale of Arab arms purchases over the past two decades. Furthermore, the comparison offers some measure of the challenge facing the Arab states should they seek to produce more of their own weaponry, since their performance would have to be set against that of the AICs, with their indigenous military industries and well-established, civilian manufacturing bases.

Unfortunately, financial details of arms deals between the Arab states (or many Third World countries for that matter) and foreign suppliers are often unavailable or incomplete. This makes any estimate of the relative share of each weapon category within total imports so imprecise as to be virtually meaningless. Even the pricing system adopted by such institutes as SIPRI, based on trend indicators

TABLE 2.1 Principal major weapons holdings of Arab states, 1973 and 1990

Type	1973	1990	Net increase	Percentage net increase	Aggregate increase[a]	Percentage aggregate increase
Land Forces:						
Tanks	6,869	20,931	14,062	205	15,962	232
APCs/MICVs/ Recce[b]	7,115	33,114	25,999	365	27,556	387
Artillery[c]	2,308	11,355	9,047	392	9,563	414
Air Forces:						
Combat a/c[d]	1,804	3,339	1,535	85	2,423	134
Transport a/c	238	434	196	82	—	—
Trainers	357	1,348	991	278	—	—
Helicopters[e]	559	1,953	1,394	249	—	—
Sea Forces (1989):						
Major Surface Combatants[f]	13	47	34	262	—	—
Submarines	12	25	13	108	—	—
FAC/Ms/Patrol[g]	208	338	130	62	206	99

Source: The Military Balance, London: IISS. (Editions for 1973–1974, 1989–1990, and 1990–1991).

Notes

[a] Total increase includes replacements for combat losses and obsolescence. Total increase has been measured for certain combat systems only.

[b] APCs/MICVs/Recce includes armoured cars and wheeled tank destroyers.

[c] Artillery includes both towed and self-propelled types, but not multiple-barrelled rocket launchers.

[d] Combat aircraft includes fighters, ground attack, reconnaissance, counter-insurgency, bombers and such support systems as tanker and EW aircraft. Also counted are trainers used in the strike role.

[e] Helicopters include attack, transport, and liaison types.

[f] Major surface combatants include destroyers, frigates, and corvettes. Figures on all naval vessels are for 1989.

[g] FAC/Ms/Patrol boats include missile- or torpedo-armed vessels generally of 80–100 tons or above. Gun-armed coastguard boats, especially of lower displacement, are not included.

rather than real market prices, is inadequate in light of the fact that actual numbers of weapons involved are often imprecise or unknown, particularly in deals involving Soviet bloc countries (up to 1989–90). Thus the best assessment is that offered by SIPRI for the entire period of 1971–85 with little chance for a more detailed breakdown: of total major weapons imports, aircraft accounted (by value) for some 52 per cent; armour and artillery for 22 per cent; guidance and radar systems for 10 per cent; missiles for 11 per cent; and ships for 5 per cent.[7] This picture, inexact as it may be, highlights the share that each main weapon category absorbs of overall imports of combat hardware.

TABLE 2.2 Comparison of Arab major weapons and those of selected members of the Warsaw Pact and NATO, 1989

	Type			
Country or Group	Tanks	Other armoured vehicles	Artillery	Combat aircraft
France	1,570	4,885	776	711
UK	1,441	4,577	550	852
GDR	2,850	5,700	1,200	367
Poland	4,050	4,400	2,025	705
Arab states	18,676	22,906	10,166	3,163

Source: The Military Balance 1988–1989, London: IISS.

Distribution According to Recipient

A further element in evaluating the relative importance of weapons imports in the Arab region is to study the distribution according to recipients. Two patterns need to be analysed: the relative share of individual Arab countries in the regional totals; and the ratio of imports to defence expenditure in each. Analysis of individual (country) shares will show whether there is a grouping of expenditure in particular countries or a coincidence of various factors such as military build-up and income levels. The imports/defence budget ratio will suggest the relative importance of hardware acquisitions in comparison with other outlays, such as construction or recurrent costs, such as maintenance and pay.

The purpose of these evaluations is not to produce precise estimates of defence expenditures, nor to analyse the impact of military expansion on development, but to obtain a general idea of the relative importance of weapons imports. It is against this background that the incentives for, and the utility of, local arms production can be assessed.

As was pointed out earlier, there are several problems in costing arms imports. Not only are official figures on total costs frequently unobtainable, but it is even more difficult to obtain accurate breakdowns of the separate cost of each category within package deals. What are the shares, for example, of the actual combat hardware, its accompanying ammunition or spares, training, and future maintenance or logistic support? Of course, not all these items may be included in certain deals, but it is often unclear whether they are or are not in specific cases.

Another problem is that the purchase price in deals may be covered by external loans or grants, or by barter deals (in exchange for oil, for example). They may also be spread out over a period of many years, so that the full cost does not appear in official defence budgets. An added complication is that the cost of weapon systems and military equipment (or services) varies widely between different suppliers, especially between Western and Soviet-bloc countries. Thus, for the same amount of money, one Arab state may buy many more tanks or combat aircraft from the Soviet Union than another Arab state might do when buying equivalent systems from the USA or France. Overall import values may thus be deceptive.

With these qualifications in mind, Table 2.3 offers details of Arab imports of major weapon systems in the period 1971–85. These figures are based on the SIPRI weighted-pricing method that suggests the price of equipment actually acquired or agreed on, regardless of the specific terms of sale or payment. This allows a better picture of the extent of arms procurements, even if it does not represent the true or exact amounts involved. At the least the figures provide a relative guide when applied across the range of Arab states. Moreover, the patterns in imports and spending related to major weapons may be assumed to be broadly similar for other military items such as small arms and support equipment, despite certain variations. Significant variations should be expected for some countries that produce their own ammunition, for example.

Several matters stand out immediately from Table 2.3. One is that there is a grouping of four principal Arab importers, each of which accounted for 15–18 per cent of the total imports in the period concerned. Together, Egypt, Iraq, Libya, and Syria bought 62 per cent of all Arab imports, worth $105bn. However, the figures are somewhat deceptive, as they overemphasise Libya's share and downplay the Saudi one. Libyan arms imports have been dropping gradually since the early 1980s, after a peak in the second half of the 1970s. On the other hand, Saudi Arabia has concluded several massive arms deals since the mid-1980s, and after the end of the Iraq-Iran war became the leading Arab importer of foreign weapons, a position it has reinforced in the wake of the Gulf crisis and war of 1990–91.

Thus the group of main Arab importers should include Saudi Arabia and exclude Libya, which had dropped back by the end of the 1980s. Another notable feature is that there is a second group of Arab states, each of which accounts for 2–4 per cent of total imports. These are Algeria, Jordan, Kuwait, Morocco, and the UAE. The remaining countries account for a descending proportion of arms imports, from 1.42 per cent downwards.

With the eruption of the second Gulf war in January 1991, all four countries in the top group of arms importers (modified to replace Libya with Saudi Arabia) have been actively involved in inter-state

TABLE 2.3 Value of Arab imports of major weapons, 1971–1990 (in $m, at constant 1985 prices)

	1971–75	1976–80	1981–85	1986–90	Total 1971–90	Percentage of Arab total in 1971–90[a]
Algeria	266	3,346	2,078	930	6,620	3.98
Bahrain	5	51	309	929	1,294	0.78
Djibouti	—	9	33	8	50	0.03
Egypt	7,376	2,918	10,182	4,717	25,193	15.13
Iraq	2,042	5,559	15,170	10,314	33,085	19.87
Jordan	562	1,908	2,363	857	5,690	3.42
Kuwait	345	1,663	1,327	487	3,822	2.30
Lebanon	96	146	480	88	810	0.49
Libya	2,804	11,725	5,721	2,247	22,497	13.51
Mauretania	9	63	17	—	89	0.05
Morocco	169	2,638	1,491	510	4,808	2.89
Oman	301	535	883	549	2,268	1.36
Qatar	30	166	934	331	1,461	0.88
Saudi Arabia	1,070	4,218	7,147	10,838	23,273	13.98
Somalia	516	268	521	43	1,348	0.81
Sudan	57	209	347	189	782	0.47
Syria	6,514	4,041	10,030	4,191	24,776	14.88
Tunisia	35	191	646	70	942	0.57
UAE	447	757	1,340	1,541	4,085	2.45
Yemen, North	8	1,016	271	63	1,358	0.82
Yemen, South	310	1,123	369	461	2,263	1.36
Totals	22,942	42,550	61,659	39,363	166,514	
Iran					17,535	
Israel					15,953[c]	
Grand Total					200,002	

Source: Based on Michael Brzoska and Thomas Ohlson, *Arms Transfers to the Third World 1971–85*, Oxford: Oxford University Press for SIPRI, 1988, Appendix 7; and *SIPRI Yearbook 1991: World Armaments and Disarmament*, Oxford: OUP for SIPRI, 1991, Table 7.4, pp. 208–11.

Notes:

[a] Percentages may not add up to 100 due to rounding.

[b] The real figure for Lebanon may be 1.08 per cent, once a $1bn debt to the USA and France is taken into account.

[c] The total for Israel includes only part of the purchases from the local arms industry.

conflict. These countries play an evident role as principal regional actors, a role each has always sought, and still seeks in the wake of the Gulf crisis, to reinforce.

In contrast, the second group of importers is marked by lower levels of conflict and ambition, although its members are not free of external threats to their national security. This has resulted in a more moderate, but still significant, pattern of military build-up. Jordan has not been involved in major inter-state combat since 1967, but nonetheless maintains combat-readiness and seeks a credible defence posture because of its proximity to Israel and buffer position between the latter and Iraq; Morocco has been engaged in a protracted low-level conflict with the Polisario guerilla front in the western Sahara since 1975 and competes with Algeria for influence; Kuwait and the UAE were threatened by the Iraq-Iran war and later became the main targets (rhetorical only in the case of the UAE) of Iraqi animosity in the lead-up to the invasion of August 1990.

This however does not mean that the remaining Arab countries have neither experienced inter-state wars nor faced external (and internal, for that matter) security threats. Somalia, for instance, fought a war with Ethiopia in 1977–78 and reached a state of total collapse in 1990–91. Oman faced an insurrection and border clashes with South Yemen in the mid-1970s, and both North and South Yemen witnessed recurrent border battles and bloody internal revolts at various times between 1977 and 1986 (before finally achieving unity in spring 1990). This commonality of security concerns is reflected in the fact that these four countries are grouped closely in terms of arms imports in 1971–85, at roughly 1–1.4 per cent.

Lebanon is difficult to place properly, partly because accurate official accounting has been lacking or else expressed in currency of constantly-shifting exchange value, and partly because of the proliferation of militias with their private, unaccounted military spending. Official figures suggest a share of 0.57 per cent of total Arab imports – low for a country suffering so much strife, but made explicable by the protracted paralysis or even disintegration of government and army since the mid-1970s. However, these figures overlook two facts: that the Lebanese Army received much free equipment during periods of reconstruction after 1976 and 1982, and that it still owes some $1bn to the USA and France in military debts.[8] In light of the latter, Lebanon's share of overall Arab procurements may be some 1.08 per cent, placing it in the preceding group.

Finally, there are five Arab countries with particularly low shares in total arms imports. Of these, two – Sudan and Tunisia – are populous, at least in comparison with other group members. The former is moreover waging an internal war, which normally would lead to higher military spending. Yet here the scarcity of financial resources plays a bigger part. Indeed, the general pattern is, not surprisingly, for fewer arms imports in smaller or poorer countries. Djibouti and Bahrain are prime examples, as is Mauretania. Qatar is highest in this group, as might be expected in light of its general security environment.

Distribution According to Supplier

The overwhelming extent to which the Arab states are dependent on outside suppliers for all major military needs is self-evident. This might appear as an argument in itself for indigenous arms production. However, by over-simplifying the undeniable reality of Arab arms dependency and by over-stating the case for reducing it through the substitution of local products for imports, such an approach can lead to distorted conclusions about where the real dependency lies and how to reduce it in practice.

Careful reading of arms-trade statistics reveals the relative importance of each supplier, in terms of volume of exports to Arab recipients, as summarised in Table 2.4. Since weapon prices vary widely from one supplier to another, there may, in fact, be greater or lesser advantage in substituting imports from each. Of course, the quality of the original product, indigenous technical ability to reproduce it, and the eventual marketability of the local copy are all additional factors to be taken into account. But acquisition patterns also show the extent to which certain types of equipment are common to several Arab armies: this suggests that local production of licensed copies or of equivalent systems might be technically and economically feasible, the more so if several countries were to co-ordinate their requests from the same supplier. Co-ordination of Arab procurements could influence a given supplier to extend licences or permit technology transfer; alternatively, diversification of sources might improve Arab prospects for obtaining such benefits.

Table 2.5 next summarises the shares, in absolute and relative terms by the use of trend-indicator prices, of the seven main world suppliers of major weapon systems to the Arab countries in 1971–90. An eighth column is devoted to imports from Third World suppliers. These are not mentioned by name, but principal among them are Brazil, Egypt, and North Korea. The other sources of arms are not included here as they occupy a minor position and exert no real strategic influence in the region. They include neutral states, such as Austria, Sweden, Switzerland, Yugoslavia, and Finland; western countries such as Belgium and Spain; and former Soviet-bloc suppliers such as Poland, Rumania, Czechoslovakia, and the GDR.

Before we analyse the significance of these figures, Table 2.6 shows the detailed breakdown of supplier shares at five-yearly intervals

TABLE 2.4 Distribution of major weapon systems by supplier (totals up to 1989)

	USSR	USA	France	UK	Italy	China	Third World	Totals
Tanks	12,990	2,077	564	1,009	36	1,560	–	18,430
Percentage	*70.48*	*11.27*	*3.06*	*5.48*	*0.20*	*8.46*		
APCs	9,929	3,893	3,289	976	–	200	3,094	22,431
Percentage	*44.26*	*17.36*	*14.66*	*4.35*	–	*0.89*	*13.79*	
Artillery	6,618	1,551	408	226	190	–	–	8,993
Percentage	*73.59*	*17.25*	*4.54*	*2.51*	*2.11*			
Combat aircraft	1,771	423	502	123	29	276	–	3,163
Percentage	*55.99*	*13.37*	*15.87*	*3.89*	*0.92*	*8.73*		

Source: The Military Balance 1988–1989, London: IISS; and Kasim Ja'far, *Mizan al-Qiwa al-Askari fi Mantaqat ash-Sharq al-Awsat* (*The Balance of Military Power in the Middle East Region 1984–1985*). Beirut: Arab Institute for Publishing and Studies, 1985.

Note:
i) Totals are greater than the sum of individual shares because other suppliers have not been mentioned.
ii) '–' means either zero or that information was not available.
iii) Germany is not mentioned as it has not exported land- or air-based weapons to the region, and its share of naval equipment has not been high enough to enter the table.
iv) Inventories are totals up to 1989.

between 1971 and 1985 (the latest date for which detailed figures are available), to obtain a more accurate idea of the patterns. The category 'other' here includes the 'neutral' states and former Soviet-bloc arms exporters.

It is immediately apparent from these tables that several Arab countries are highly dependent on a single external supplier. This is most obvious in the case of importers from the USSR, which was the major source of arms for South Yemen (98.98 per cent), Syria (96.67 per cent), Algeria (80.36 per cent), Libya (73.32 per cent), North Yemen (65.54 per cent), and Iraq (64.83 per cent). On the other side, both Saudi Arabia and Jordan have relied primarily on American arms, which represented 54.85 and 56.84 per cent of all their imports, respectively. A reverse feature is that several Arab countries received no, or only negligible, military imports from one or other superpower in 1971–90. Most notably, the USSR provided virtually no arms to eight Arab states and only 0.7–1.0 per cent of imports purchased by five others. Only Egypt and Somalia stand in a medium position, having relied on the USSR for some 30 and 42 per cent of their arms imports respectively in the period under survey.

TABLE 2.5 Distribution of major weapon imports in Arab countries by supplier, 1971–1990 (in \$m, at constant 1985 prices, percentages of total imports in italics)

	USSR	USA	France	UK	Germany	Italy	China	Third World
Algeria	5,320	532	416	28	59	—	54	6
	80.36	*8.04*	*6.28*	*0.42*	*0.89*		*0.82*	*0.09*
Bahrain	—	759	138	20	358	3	—	3
		58.65	*10.66*	*1.55*	*27.67*	*0.23*		
Djibouti	—	—	38	—	—	—	—	12
			76.00					*24.00*
Egypt	7,460	11,067	2,756	591	11	423	1,748	292
	29.61	*43.93*	*10.94*	*2.35*	*0.04*	*1.68*	*6.94*	*0.03*
Iraq	21,450	389	5,317	—	34	921	1,291	3,247
	64.83	*1.18*	*16.07*		*0.10*	*2.78*	*3.90*	*9.81*
Jordan	481	3,234	826	478	—	—	—	433
	8.45	*56.84*	*14.52*	*8.40*				*7.61*
Kuwait	375	1,372	1,106	511	292	—		40
	9.81	*35.90*	*28.94*	*13.37*	*7.64*			
Lebanon	6	383	211	85	—	11	—	111
	0.74	*47.28*	*26.05*	*10.49*		*1.36*		*13.70*
Libya	16,494	800	2,685	56	—	1,674	—	929
	73.32	*3.56*	*11.93*	*0.25*		*7.44*		*4.13*
Mauretania	5	4	17	23	—	—	—	—
	5.62	*4.49*	*19.10*	*25.84*				
Morocco	—	1,641	2,117	15	15	75	—	491
		34.13	*44.03*	*0.31*	*0.31*	*1.56*		*10.21*
Oman	—	249	254	1,641	—	9	9	86
		10.98	*11.20*	*72.35*		*0.40*	*0.40*	*3.79*
Qatar	—	—	1,285	165	—	—	—	11
			87.95	*11.29*				*0.75*
Saudi Arabia	12,766	5,570	2,988	—	272	1,440	236	—
		54.85	*23.93*	*12.84*	*1.17*	*6.19*	*1.01*	
Somalia	562	337	3	—	—	230	139	85
	41.69	*25.00*	*0.22*			*17.06*	*10.31*	*6.31*
Sudan	10	141	46	3	25	8	91	249
	1.28	*18.03*	*5.88*	*0.38*	*3.20*	*1.02*	*11.64*	*31.84*
Syria	23,950	—	81	—	—	—	—	524
	96.67		*0.33*					*2.11*
Tunisia	—	417	360	10	—	75	4	19
		44.27	*38.22*	*1.06*		*7.96*	*0.42*	*2.02*
UAE	—	604	2,346	347	428	150	—	100
		14.79	*57.43*	*8.49*	*10.48*	*3.67*		*2.45*
Yemen, North	890	399	—	—	—	1	35	22
	65.54	*29.38*				*0.07*	*2.58*	*1.62*
Yemen, South	2,240	—	—	—	—	—	—	—
	98.98							
Totals	79,243	35,094	25,572	6,961	1,222	3,852	4,811	9,759[a]
	47.59	*21.08*	*15.38*	*4.18*	*0.73*	*2.31*	*2.89*	*5.86*
Grand Total	166,514							

Source: Based on Michael Brzoska and Thomas Ohlson, *Arms Transfers to the Third World 1971–85*, Oxford: Oxford University Press for SIPRI, 1988, Appendix 7; and *SIPRI Yearbook 1991: World Armaments and Disarmament*, Oxford: OUP for SIPRI, 1991, Table 7.4, pp. 208–11.

Notes:

i) The '—' sign means either zero or that information is not available.

ii) Deliveries only are included, not value of systems under contract but not actually delivered.

[a] The total number under the Third World heading has been increased by \$2,863m to include imports from other countries not cited in the table.

TABLE 2.6 Breakdown of supplier shares to the Arab countries by five-year periods, 1971–1990 (in $m, at constant 1985 prices)

Country or group	1971–75	1976–80	1981–85	1986–90	Overall 1971–90
USSR	83.0	43.4	36.7	36.1	47.6
USA	9.4	29.6	28.0	21.1	21.1
France	4.5	14.7	17.5	16.8	15.4
UK	1.0	4.3	4.1	8.9	4.2
West Germany	na	na	1.1	1.3	0.7
Italy	na	0.3	2.1	1.2	2.3
China	na	1.6	3.4	6.7	2.9
Third World	na	1.6	3.8		
Other	na	1.6	3.2	7.9	5.9

Source: Based on Michael Brzoska and Thomas Ohlson, *Arms Transfers to the Third World 1971–85*, Oxford: Oxford University Press for SIPRI, 1988, Appendix 7; and *SIPRI Yearbook 1991: World Armaments and Disarmament*, Oxford: OUP for SIPRI, 1991, Table 7.4, pp. 208–11.

Notes:
i) na = negligible share.
ii) Percentages may not add up to exactly 100, due to rounding.

Interestingly, the USA enjoyed a substantially different situation. Only six Arab countries bought no American major weapons: of that number two were pro-Western – Qatar and Djibouti – reducing the real exceptions to only four: Syria, South Yemen, Libya and Iraq (the latter two have bought negligible amounts of US weapons or dual-use items). And of the remaining 15 states, six (Morocco, Egypt, Kuwait, Tunisia, Bahrain, and Lebanon) received 35 to 59 per cent of their arms from the USA in 1971–90, a considerable share. Another sizeable group purchased smaller, but nonetheless appreciable, amounts of American weapons: the UAE, Sudan, Somalia, and North Yemen all lie in the 15–30 per cent band. And finally, at the bottom end of the scale Libya, Mauretania, Algeria, and Oman turned to the USA for 1–11 per cent of their major weapons.[9]

The foregoing suggests that the USA has consistently enjoyed a broader and more even spread in its arms exports and military ties to the Middle East than the USSR. This is a pattern, moreover, that is likely to deepen in the wake of the second Gulf war as the USA consolidates its strategic position in the region and the USSR reviews the quality of many of its weapon systems. Similarly, France enjoyed a wide spread in its exports in the same period, supplying all but two Arab countries (the two Yemens) and establishing itself as the third main supplier of arms to the Arabs. Its sales were equal to half the American share, and four times greater than its nearest rivals. Though responsible for far fewer exports to the region, both Italy and Britain have supplied an overwhelming majority of Arab states, as have the Third World weapons exporters/producers as a group.

Overall, all the tables show that the USSR holds a leading position as an exporter to the Arab region in 1971–90, whether expressed as a share of major weapons systems (Table 2.4) or relative value (Table 2.5). However, its relative share of total regional imports has declined considerably over time. The USA occupies second place in value and in most weapons categories except combat aircraft, where it cedes to France. The latter holds a close third place generally in weapons inventories, and a good second place in terms of value. Although there is no credible way of relating the overall values in Table 2.5 to the weapon totals in Table 2.4 (since the latter are composed of different types that cannot simply be added up), a quick comparison shows that the share of income accruing to the USA and Italy was higher than the share of any category of Soviet weapons exports to the Arabs. No such comment can be made about any of the other suppliers with any degree of reliability. For what it is worth, this observation serves merely to confirm the widely-accepted fact that Soviet equipment is generally cheaper per unit than Western equivalents.

However, although they are valid at the most general level, these initial impressions are somewhat misleading. Most importantly, they apply to aggregate totals and shares over the entire period of 1971–90. But the real share of each supplier actually changed widely within that period, and the total pattern of Arab arms dependency changed considerably. Thus by the last five-year period of 1986–90, the overall share of Soviet weapons had dropped by 47 points since 1971. In some Arab countries, the share of each supplier dropped sharply, from period to period. Soviet supplies also dropped by 27 points to 52 per cent in Algeria, 10 points to 61 per cent in Libya, and 11 points to 55 per cent in Iraq. These drops are measured against the average shares for 1971–90 in Table 2.5, and not against import levels in previous years, which would have shown a much steeper decline. After 1981, Jordan's imports from the USA dropped by 31 points, to 33 per cent below the overall average, with the share of Arab states declining by 7–8 per cent generally.

On the other hand, reliance on Soviet arms remained stable in Syria and South Yemen and even increased sharply in North Yemen. As regards American sales, Saudi and Moroccan purchases of American weapons remained constant in relative terms, while five other countries witnessed important rises in 1981–85: 15 points up to 24 per cent in Algeria, 27 points up to 64 per cent in Egypt, 23 points to 59 per cent in Somalia and 45 per cent in Sudan, and 16 points to 38 per cent in the UAE. (Four out of these five countries used to be heavily dependent on Soviet weapons in the 1960s and

1970s; no equivalent shift has occurred from Western equipment to Soviet items, though both Kuwait and Jordan considered and even implemented partial shifts in the 1980s.) There have been lesser rises of 4–8 per cent in five other Arab states importing from the USA. Of course, a word of caution is necessary: especially in the smaller countries, with small defence expenditures, a single transaction may alter completely the apparent import pattern, though the impact on larger countries with large expenditures over many years is much reduced.

To draw any major conclusions from these observations would be risky, however. For one thing, although relative shares have declined in some cases, absolute values of imports from any particular supplier may not have dropped, as the base for calculation (total military expenditure in any one five-year period) rose considerably in later years for most countries. Thus Iraqi purchases from the USSR dropped from 85 per cent of all major weapons imports in 1976–80 to 55 per cent in 1981–85, but rose in value from $4,725m to $8,343m.[10] What is worth noting, however, is the general trend towards reducing dependence on a single source of supply. This was coupled with a related trend towards multiple diversification, rather than a simple switch of dependency from one supplier to another or to two (rather than several) suppliers. France was most often the third alternative. At the same time, both China and the group of Third World suppliers (principally Egypt and Brazil in the Arab case) generally increased their share in the market in the 1980s, though the pattern of growth was not consistent.

In observing these trends, several further remarks can be made. One is the massive impact of specific events, such as the Iran-Iraq War. Iraqi imports alone (not counting Iran nor increased spending in other Gulf states as a result of the rise in tension) were worth $4,725m to the USSR and $3,337m to France in 1981–85, or 13 per cent of all Soviet exports and 25 per cent of all French exports to the Arabs in 1981–85. The end of that conflict, and the need to cope with declining oil revenues and with the enormous costs of economic reconstruction and debt-servicing, seemed to herald reductions in military spending, if only by *force majeure*.

Whether or not the second Gulf war will significantly alter the long-term trends is not a foregone conclusion. It may trigger a new escalatory spiral, even with Iraq under an arms embargo, as readily as stimulate serious regional arms controls. The tendency to update weapon systems and maintain military development is driven by a number of powerful and lasting security, political, and bureaucratic (institutional) factors, but it may be modified and regulated. If conflict is, in fact, brought under control and regional security enhanced, the procurement trend may slow down. In that case, large arms deals may leave an even greater impression on the overall trend than before. Prominent examples are the Yamamah-1 and Yamamah-2 deals, worth $25bn, concluded by Saudi Arabia with Britain in 1985 and 1988, and the $20bn request made by the Kingdom to the USA in 1990.[11] Among the largest such contracts in history, they will alter the balance (and weigh the statistics) of the Saudi-USA, Saudi-British, and Arab-Western arms trade for years to come.

Implications

This last example has particular significance for the future of Third World exporters and producers of military equipment. Simply put, Arab states still tend almost automatically to turn to the AICs for major imports of hardware and technology (in the form, for instance, of training, maintenance, engineering, design, and construction services). This trend is likely to grow in peacetime rather than decline, for several reasons. Those Third World countries that did not produce armaments but merely re-exported them enjoyed their main opportunity to do so during war-time.[12] Secondly, most indigenous Third-World military products imported by the Arabs have been ammunition, small arms and rocket artillery, and light, wheeled armoured vehicles, which are in greatest demand during active conflicts. Finally, peacetime investment, by the same token, is greater in expensive systems than in ammunition, small arms, and spare parts, so redirecting trade towards the high-tech producers. Given this reality, can Arab defence industries compete, and in what areas of production?

3

Assessing Arab Defence Spending and Arms Imports

Introduction

In focusing on patterns of Arab arms imports, the primary purpose of Chapter 2 was to emphasise their importance, both financially and militarily. A secondary aim was to note the durability both of the tendency to import arms and of military expenditure in the Arab region – patterns that are likely to endure despite some fluctuation. In a broad way, therefore, the preceding chapter indicated the need for the Arab states to participate more actively in providing their own hardware requirements. It also showed that, because such major resources are involved in purchasing foreign weapon systems, to embark on local military industrialisation might appear a feasible and tempting policy.

Subsequent chapters offer a detailed discussion of the technical, industrial, and economic requirements of indigenous Arab defence production. Here the aim is to assess the general need for local military industrialisation, by presenting a picture of its broad financial and economic determinants. In practical terms this means evaluating the overall volume of defence spending and comparing the proportion of arms imports to it, since these two items are the most basic factors that motivate moves towards local arms production. In order to provide a preliminary indication of the areas of greatest need for local industrial activity, the following section comments on the patterns of Arab hardware acquisitions. The final section offers an abbreviated survey of the economic, financial, and demographic indicators that affect the ability of individual Arab states to undertake indigenous arms production. Of course, theoretical ability is no guarantee of actual capability, but the purpose here is merely to provide the background to subsequent chapters.

Methodological Notes of Caution

Before we move into a substantive discussion of the preceding issues, several comments need to be made about the availability of official data on Arab defence expenditures. The issue of information presents a particular difficulty in the Arab region, both for government agencies and for independent analysts and researchers. However, the aim here is neither to debate the reasons for this situation nor to speculate on its future consequences – though there can be no doubt that the Arab countries will face increasing difficulty in competing in world markets or in developing their economies and infrastructures in the era of the 'information revolution'. Rather, the purpose of this section is to stress a number of cautionary notes, to be kept in mind whenever data are used throughout this book.

The first problem in calculating defence expenditure relates to official defence budgets. It is often unclear in Arab countries if whether they include all items of expenditure or only some. In certain cases, the budget announced may be that of the defence ministry and relate to recurrent costs (pay and maintenance), while the army has a separate budget for acquisitions, which is not published. In some countries, such as Algeria, the budget is specifically stated not to include equipment, while in others, such as Tunisia, all components are specifically included.

A second complication is that several Arab countries – such as Jordan and Syria, or Iraq during the Iran-Iraq war – have received external assistance from other Arab governments, which is not always declared or categorised. In Jordan, for example, grants or loans made for specific weapons purchases are not included in the official defence budget. Conversely, it is not clear how external budgetary support is accounted for: does it form part of the defence budget, or is it considered as general support that allows more local revenue to be allocated to defence? Similar questions apply in other cases. Does Arab aid appear in the Syrian defence budget (whatever its purpose may be, support or procurement), or is it paid directly to arms suppliers on behalf of Syria? Has Arab aid for Iraq gone straight into paying military salaries or buying arms or into servicing the Iraqi debt to foreign lenders, whose loans were made to finance weapon imports?

A similar situation arises with respect to foreign loans to Arab countries for military acquisitions. Major examples are the American Foreign Military Sales and the Military Assistance Program, which support most sales to friendly Arab states. To begin with, are such loans included at all within official defence expenditure figures, or are they left out and treated simply as external public debts? And when accounted for in any way, do the Arab governments concerned include the full value of the annual American allocation, or do they spread it out over the full period of the loan, showing a much smaller annual amount?

A consequent question is how the debts incurred for military spending are accounted for. Nearly half the Arab countries have accumulated large debts to foreign nations, substantial parts of which are due to loans to buy weapons and other military needs. Indeed, debts due to military borrowing may be as high as 20 per cent of the total in Egypt (before the 1990 write-off of American military debt); just how much of the estimated $80bn Iraqi debt resulting from the war with Iran was due to the arms purchases is not known, but it must have been a considerable share. Where do loans appear in government figures, and where in the budget is the interest paid on loans accounted for? Debt servicing is most probably not included at all within defence budgets – though the capital might be – and interest paid on military loans is not specifically designated as defence related within statements by the treasury, central bank, or other official body.

An additional, complicating factor in calculating Arab defence spending is that a number of countries have resorted in the past to the exchange of goods – especially raw materials and primaries, such as oil and cotton – in return for arms shipments. Egypt did this for many years with the USSR, while Saudi Arabia (and several other oil-exporters including Iraq and Libya) has bartered oil for up to 50 per cent of the value of arms deals, such as the Yamamah-1 contract signed with Britain in 1986. Indeed, Saudi barters have accounted nominally for up to 85 per cent of the value of military construction contracts. In virtually all cases, counter-trade is not accounted for in official military statistics and defence budgets, though it can evidently have major economic significance.

A further complication is how to account for local arms production in those Arab countries where this is relevant. Most concerned is, of course, Egypt, though Iraq emerged in the mid-1980s as a growing industrial force. Other Arab countries, such as Saudi Arabia, Algeria, and Syria, also produce limited quantities of military items. Has the potentially massive investment involved in military industry – for machinery, buildings, land, materials and sub-systems, and R&D – been debited to the defence budget? In Egypt the Defence Ministry directs the industry but holds no financial control; part of the cost is budgeted to the Ministry of Industry. Furthermore, earnings from sales by government-owned factories may accrue to the army: is this income accounted for within the overall defence budget or is it extra-budgetary; is it spent on army purchases; and if so under what budgetary heading are they accounted for?

In all cases, although official Arab defence budgets may offer a total for non-equipment costs (including maintenance, training, and construction) or for pay, none offers more detailed breakdowns. Thus the component spent on repair and maintenance is unknown, as is that for construction, training, or supplies. (Though such a breakdown may be available for contracts signed with Western suppliers.) Much of this local expenditure is moreover disguised within the budgets of other ministries, obscuring the volume of overall defence expenditure as well as its categories. Thus construction or bridges building and roads paving might be assumed in the records to be at the cost of the ministry of construction or public works.

Connected to the problem of statistical breakdowns is the fact that different countries assign different degrees of importance to items such as pay. Thus Egypt or Syria may keep that budget component low by paying low salaries to career personnel and even lower wages to conscripts, while Saudi Arabia offers very high salaries to its all-professional army.[1] Besides making comparison or projection between countries risky, this discrepancy also indicates that the degree of distortion in data can be even higher than commonly thought. That is, the higher the proportion of arms procurement to overall (real) spending, the greater the amount that is often not included in official budgets or that is subsidised, 'spread out' in time, or otherwise disguised.

Defence Spending and the Ratio of Imports

With these cautionary notes in mind, it is possible to draw a broad outline of both the volume of Arab military expenditure and the importance of arms imports within it. This is based on official statistics or on estimates by Western and international agencies using compiled information on known arms deals for major weapon systems. Table 3.1 compares aggregate arms imports with total military expenditure for the Arab countries in three periods (1976–80, 1981–85 and 1986–90).

In order to assess the significance of the figures in Table 3.1, the previous cautionary notes need to be applied. As noted in the table, information is inadequate or absent in the case of certain countries, especially Qatar and Libya (and Lebanon since 1984). The same applies, to a lesser degree, to Bahrain, Mauretania, Somalia, Sudan, UAE, and the two Yemens. The impact of this drawback is softened, however, by the fact that there are reasonably adequate data on all the major Arab defence spenders and arms importers (except Libya).

A greater cause for concern, though, are other flaws in the information. An important one is that the SIPRI import data used here refer only to major weapon systems, but not to small and medium arms, non-combat support equipment, and *matériel* for the internal

TABLE 3.1 Comparison of Arab arms imports and c

Country	Defence budgets		
	1976–80	1981–85	1986–90
Algeria[a]	2,465	4,530	5,168
	5,811	*6,608*	*6,098*
Bahrain	166[c]	792[c]	663
Djibouti	na	113	126[e]
Egypt	16,379	15,567	21,360
Iraq[g]	10,960	49,777	60,349
Jordan[a]	1,451	2,465	3,259
	3,359	*4,828*	*4,116*
Kuwait[a]	4,797	7,625	7,304
	6,460	*8,952*	*7,791*
Lebanon	882	2,052[c]	326[e]
Libya	1,179[c]	1,266[d]	5,630[e]
Mauretania	na	175[c]	—
Morocco	2,877	4,623	5,297
Oman	3,559	9,212	7,374
Qatar	121[d]	1,062[c]	—
Saudi Arabia	58,811	117,175	75,660
Somalia	220[c]	652	147[c]
Sudan	726[c]	1,360	2,127
Syria	9,273	14,452	12,361
Tunisia[a]	739	2,359	2,501
	930	*3,005*	*2,571*
UAE	2,611[c]	9,247[e]	8,110
Yemen, North	568	2,197[e]	1,525[c]
Yemen, South	283[c]	685[e]	427[d]

Source: For defence budgets, *The Military Balance*, London: IISS. For arm
Oxford: Oxford University Press, 1988, Appendix 7; and *SIPRI Yearboo*
1991, Table 7.4, pp. 208–11.

Notes:
i) Most figures exclude internal security and foreign military assistance.
a) For these countries defence budgets do not include equipment, and so ar
defence budgets plus imports of major weapons, but leaves out all other i
defence spending in italics.
b) Ratio of imports to defence budget (milex); percentages are rounded-off
c) Covers 3 years of period.
d) Covers 2 years of period.
e) Covers 4 years of period.
f) Ratio is calculated by taking average yearly expenditure and imports.
g) Estimates of Iraqi defence spending since 1980 are based on Gulf bankir

›ending, 1976–90 (in $m, at constant 1985 prices)

◂Iajor arms imports		Import/milex (percentage)[b]		
1981–85	1986–90	1976–80	1981–85	1986–90
2,078	930	58	31	15
309	929	16	19	42
33	8	na	29	5
10,182	4,717	18	65	18
15,170	10,314	51	30	15
2,363	857	57	49	21
1,327	487	26	15	6
480	88	17	19	22
5,721	2,247	86	64	32
17	—	na	10	—
1,491	510	68–92[f]	26–32[f]	10
883	549	15	10	7
934	331	35–55[f]	34–53[f]	—
7,147	10,838	7	7	14
521	43	42–73	44–80	18
347	189	17	26	9
10,030	4,191	44	69	34
646	70	20	27	3
1,340	1,541	17	12	19
271	63	65	10	2
369	461	76	30	13

, Brzoska and Ohlson, *Arms Transfers to the Third World 1971–1985*, *'orld Armaments and Disarmaments*, Oxford: Oxford University Press,

ıate total for defence spending is placed in italics. This consists of stated ·gories. In all these cases, the import/milex ratio is based on the aggregate security forces. Thus the real level of arms imports is substantially higher than suggested. It is raised even further when follow-on services are taken into account. The latter include repairs and maintenance, spare parts, training and advice, and construction – some, but by no means all of which are included in import figures. In Saudi Arabia, for instance, purchases of hardware may represent only one-quarter of total foreign contracts, with other services not necessarily included in import figures.[2] One implication of the foregoing is that in countries like Algeria and Tunisia, whose publicised defence budgets cover only recurrent costs but not equipment, the total volume of imports and thus of total defence spending is greater than shown in Table 3.1. Elsewhere, the UAE presents a special case, as its published federal budgets and expenditure do not include any additional purchases made by its individual member states, thus understating the total volume of imports to the region.

Another critical issue is that Arab defence spending is seriously under-reported by governments through a variety of mechanisms. Most common is the distribution of certain costs among other ministries and budget categorisations. Internal security is the most obvious case in point, as few Arab countries include it in their defence budgets. And because it pays for them, part of military imports are attributed to the treasury rather than the ministry of defence or the armed forces. The budget of the Saudi Military Academy is subsumed under that of the Royal Court, for instance, while Army medical costs are budgeted under the Ministry of Health. At a different level, the cost of establishing and running local defence industries is also excluded from official and unofficial estimates of military expenditure in Egypt and Iraq. (It is unclear, though probable, if the same applies to Saudi Arabia or other states.)

Foreign military assistance and loans pose another costing problem. Egypt, for example, does not include American aid in its statistics. Nor does it count under the heading of defence the cost of servicing debts incurred through military spending (mainly to the USA, the USSR, and France), although both the capital repayments and interest are in fact part of defence spending. The same applies to Jordan and Morocco especially, and to Somalia, the two Yemens (pre-1990), Tunisia, and Lebanon (the latter owes some $1bn to the USA and France for hardware bought in 1983–84). Bahrain and Qatar benefit from a $1bn subsidy fund set up by the GCC, which is spread out in grant form over 1984–94. Syria owes some $7bn to the Soviet bloc for arms purchases, but although it accounts for capital repayments in the defence budget, it does not include servicing. In the case of Iraq, it is unclear how much of its $75–80bn war debt is due to defence imports, or whether that part is already included in external estimates of its total military spending during the war.

A final complicating factor is the fact that most Arab oil-exporters have used barter deals to finance a large part of their defence-related imports. Up to 50 per cent of Saudi military construction contracts with foreign companies have been funded in this way, as were some 70 per cent of Libyan arms purchases in the mid-1980s. Thus although the figures offered for imports of major weapons might not change, the real volume of defence spending might be seriously under-reported.

The implications of the above for any assessment of real Arab import and defence spending levels are as follows. First, total military imports are made appreciably larger when non-major items are included. The same can be said of imports of non-hardware services. In the case of Saudi Arabia, this could mean a fourfold jump in the total value of defence-related imports, with a corresponding increase of the import/military expenditure ratio from 7 to 28 per cent. Second, the overall cost of defence *and* security is substantially higher than estimated, once all related items and allocations are accounted for, wherever they may be hidden within official budgets. This is even more obvious when external assistance is included, which increases the national burden, and when the exchange of non-cash resources (barter) is implemented in return for hardware and services. The latter items alone would result in at least a 35 per cent increase over the official Egyptian defence budget, for example. In general, the arithmetical impact is likely to be least in countries with low import and defence spending levels, and greatest in those that have high levels or that resort heavily to barter.

Observations on Arab Weapons Acquisitions

The figures presented in the preceding section and in Chapter 2 confirm an exceptionally high level of arms imports to the Arab countries, with an emphasis on major weapon systems. This indicates both massive financial expenditure and what are perceived locally as pressing security and defence needs. Whatever the incentives may be, the idea of indigenous military industrialisation would seem entirely logical and feasible, given the extent of the market and of the human and material resources allocated to security/defence.[3]

The Feasibility of Arab Weapons Acquisitions

To embark on such a large-scale venture as arms production requires a critical look at the patterns of Arab military development, though. Specifically, the question must be asked whether the volume of arms imports is commensurate with Arab capabilities for both absorption and deployment. Modern weapon systems are highly complex technologically, and require special operator skills and inordinate amounts of maintenance and follow-on support (requiring yet further qualified manpower). Furthermore, the increasing movement towards state-of-the-art PGMs (especially 'fire-and-forget' munitions), stand-off weapons, and long-range guided systems or cluster and dispenser warheads is leading to a 'saturated' battlefield, theoretically requiring fewer main weapons platforms. Are the Arabs acquiring more hardware than they can use, therefore, and how much of it can they, or should they, seek to produce locally?

In order to provide a response, Table 3.2 shows a rough comparison of the manpower-to-equipment ratios. This is by no means an accurate and decisive criterion, but it does indicate generally whether each Arab country is acquiring as much as, or more than, it can usefully absorb. As an added measure, the same ratios are calculated for Israel, in order to offer a more precise and telling comparison.

It must be stressed, once again, that these figures and ratios are by no means an exact indicator. Equipment types, and therefore operator and maintenance requirements, can vary widely. Moreover, in some Arab countries certain back-up skills and follow-on services are provided by foreign personnel. An army with a large foot infantry component will show a high manpower-to-equipment ratio although its armoured arm may, in fact, have a low one. And of course, different armies have different operational objectives and technical requirements: armies waging counter-insurgency or mountain wars – such as those of Morocco, Sudan, and Oman – have less need for major weapon systems and their support services. This helps to explain the high manpower-to-equipment ratios in these countries, which is also characteristic of poorer or less-developed ones, such as Somalia and Mauretania.

Nonetheless, it is equally obvious that several important Arab countries possess amounts of military hardware that they cannot comfortably operate and maintain. Indeed, if the foregoing comment applies in peacetime, then it is of paramount importance in combat situations in which foreign personnel are unlikely to work and which impose high attrition rates owing to mechanical and other failure.

This observation becomes even more pressing when Arab ratios are compared with the Israeli ones: 56 men per AFV, 41 men per AFV or tank, 56 men per combat aircraft. The lack of reasonable proportion between manpower and equipment is most evident in the oil-exporting countries, especially Saudi Arabia, Libya, and Kuwait. Notably, however, the same applies (albeit to a lesser degree) to both Syria and Jordan, which are principal confrontation states against Israel historically. With the one exception of Syrian manpower per AFV, all ground force ratios are substantially lower than for Israel, apparently implying significantly greater ability to field AFVs and

TABLE 3.2 Manpower-to-equipment ratios in the Arab and Israeli armies, 1989

Country	Army manpower (000)	Tanks and other AFVs	Other AFVs	Manpower: tanks and other AFVs	Manpower: other AFVs	Air force manpower (000)	Combat aircraft	Manpower: aircraft
Algeria	120	2,730	1,780	44	67	12	299	40
Bahrain	2.3	195	141	12	16	0.45	12	38
Djibouti	2.87	63	63	46	46	0.1	0	na
Egypt	320	6,135	3,695	52	87	30[b]	517	58[b]
Iraq	955	14,600	9,000[a]	65	106	40[b]	513	78[b]
Jordan	74	2,565	1,415	29	52	11	111	99
Kuwait	16	945	670	17	24	2.2[c]	36	61
Lebanon[d]	21	527	390	40	54	0.8	2	400
Libya	55	3,910	1,930	14	28	22[b]	515	43
Mauretania	10.4	120	120	87	87	0.25	5	50
Morocco	170	1,652	1,378	103	123	16	109	147
Oman	20[c]	98	29	204	690	3	63	48
Qatar	6	224	200	27	30	0.3	13	23
Saudi Arb.	38	2,600	2,050	15	18	16.5	179	92
Somalia	61.3	823	520	74	118	2.5	63	40
Sudan	65	602	357	108	182	6	45	133
Syria	300	8,350	4,300	36	70	40	499	80
Tunisia	30	427	255	70	118	3.5	43	81
UAE	40	885	678	45	59	1.5	61	25
Yemen, North	35	1,204	540	29	65	1	123	8
Yemen, South	24	1,085	605	22	40	2.5	114	22
Israel[e]	598	14,574	10,780	41	56	37	676	56

Source: Based on *The Military Balance 1989–1990*, London: IISS, 1989.

Notes:

i) AFVs do not include self-propelled artillery or air-defence systems.
ii) Manpower-to-equipment ratios rounded to nearest whole integer.
iii) Combat aircraft do not include armed helicopters.

[a] 1985 figure.
[b] Includes air defence units, and so manpower-to-aircraft ratios are in fact substantially lower.
[c] Does not include foreign personnel.
[d] Figures inaccurate due to break-up of army and proliferation of militias.
[e] Israeli manpower figures after mobilisation.

tanks. In reality, this is not the case, which suggests that Arab armoured combat performance will be adversely affected in practice.

A final qualifying remark about the use of manpower-to-equipment ratios is that the real tests of human military skills in this context are turn-around times, retrieval rates of damaged equipment, and sortie rates. These refer to the amount of time consumed in resupplying and otherwise preparing weapon systems to return to battle; the ability to repair damaged hardware and return it to service, on the field and within hours, or at rear depots and within days; and the number of times combat aircraft can be put into the air in a single day or period of time (a function of turn-around time). It is these and other factors that really determine whether a given manpower-to-equipment ratio is sufficient. That is, although most Arab air forces have far higher ratios than the Israeli Air Force, the latter can generate higher sortie rates, owing to superior training, scientific background, and organisation. Conversely, an armed force that is overloaded with equipment is likely to become increasingly inefficient as the system suffers breakdowns.

Aspects of Commonality

An underlying thrust of the preceding section is that the Arab countries need to invest more human and financial resources, and time, in the wide variety of follow-on services required to keep modern armies and their sophisticated hardware in operation. The core element of such a process is human development, as it is the preparation of skilled manpower that is the key to military development.

A further implication is that the Arab countries might be well advised to reduce the level of their arms imports, and to focus either on less technologically-complex hardware or else on special sub-fields of modern military technology, rather than continuously acquiring the most advanced major weapon systems possible. This does not mean being satisfied with equipment that is inadequate for Arab security/defence needs, nor that the Arab armies should not strive to raise their technical capabilities. But it does mean that a more effective approach to achieve a long-term indigenous defensive ability – that is autonomously generated – might be to build up human skills by the use of more appropriate equipment, which is moreover cheaper. That this need not result in the use of low-quality weapon systems is demonstrated by the Kuwaiti decision to buy modern, Yugoslav-made T-72 tanks, or by the Saudi option of buying Brazilian Osorio tanks (not followed up, however), instead of far more costly British Challengers and American M-1s.

As crucial in the context of this discussion, though, is to note the degree to which both the types of military hardware and the problems of operating and maintaining them are shared among the Arab countries. In the first case, as Table 2.4 showed, there is incontestable Arab dependence on a few foreign suppliers of major weapon systems. The USSR had supplied 70 per cent of all tanks in Arab service at the end of the 1980s, 44 per cent of the APCs, 74 per cent of gun artillery, and 56 per cent of combat aircraft. The same shares for the USA stood at 11, 17, 17, and 13 per cent, respectively. Of all other sources, France accounted for 15 per cent of all APCs and 16 per cent of all combat aircraft, with another 14 per cent of APCs coming from the Third World (mainly Brazil), and 8 and 9 per cent of tanks and combat aircraft supplied by China.

Furthermore, the weapon types that make up these shares are relatively consistent throughout the region. That is, most Soviet-supplied Arab armies field T-54/55, T-62 and now T-72 tanks, BMP-1 APCs, 122/130mm guns and 122mm MRL systems, and MiG-21 and MiG-23/27 (and now MiG-29) combat aircraft to name but a few. Conversely, M-48 and M-60 tanks (and now M-1s), M-113 APCs, M-109 155mm SP howitzers, and F-5 and F-16 fighters are common among US-supplied countries (though aircraft commonality is lower). Similar remarks might apply to French *Mirage* F-1s and *Mirage*-2000s, or German *Lurssen* TNC-45 missile boats and Italian *Asaad* frigates. (The preceding should not be taken to deny the equally notable degree of diversification of types and sources, which has its own advantages and disadvantages, such as greater expense and complexity in stockpiling spare parts and munitions.)

This generally high degree of commonality suggests that it would be feasible – both financially and technically – for the Arab states to develop an intergrated or at least complementary services infrastructure and/or production base. This would allow maximal exchange of experience and skills, as well as fuller, more prolonged use of facilities and qualified manpower, thereby achieving more economical returns. A variety of specific arrangements could be developed, involving greater or lesser degrees of integration, but the core issue is to note the large scope for, and potential rewards of deliberately reinforcing standardisation of equipment, technical services, and operating procedures; and of utilising such commonality to encourage inter-Arab co-operation and offer improved circumstances for indigenous military industrialisation.

General Determinants of Productive Capability

Although a full discussion of the determinants of indigenous military industrialisation, with particular reference to Arab capabilities, is presented in Part Four, it is useful at this point to offer a summarised overview of some of the relevant basic financial and economic indicators. These are: each Arab country's GNP, per capita GNP, the ratio of military expenditure to GNP, population size, and total external debt.

Although several other essential factors help to determine a country's industrial ability – not least of which are its existing industrial base, economic infrastructure, foreign exchange reserves, state revenues, and scientific/technical manpower pool – the indicators cited serve as a general pointer to the potential ability of the local (national) economy to support military industrialisation, to the possible constraints posed by indebtedness, and to the potential manpower pool (for both industrial and military development) which in turn suggests the size of potential markets for local defence products. By these standards, it is plausible to argue that it is the Arab countries imbued with large GNP and populations, and which also tend to relatively high levels of defence/security spending, that are the most likely to embark on indigenous defence production and most able to benefit from it. They are more able to provide the military industrial sector with manpower and subsidiary inputs (raw materials and components), and have larger absorptive capacity for the type and technological level of the hardware they might produce locally. This is in contrast to most oil-producers, who have the surplus funds but lack the industrial manpower and local military market to make arms production really feasible. By way of contrast, the remaining Arab countries are generally either too small or too poor to set up local arms industries on a substantial scale, though this need not prevent them from developing special areas of industrial activity, such as small arms munitions or even electronics (especially if this fits within their civilian industrial strategy).

Conclusion

In conclusion, it is well worth re-emphasising the relevance of the discussion of the impact of arms imports to the issue of military industrialisation. Ultimately the purpose is to evaluate the need for indigenous arms production, and, more specifically, to arrive at conclusions as to which spheres and types of local production are most needed. A simple method is simply to measure the full volume of hardware acquisitions, in which imports occupy a predominant place. But this should not be taken to mean that local industrialisation efforts must necessarily aim at total import-substitution, or even that a main effort must be made to produce major weapon systems – given that the latter may indeed be the most effective and sensible option. Nor does the preceding mean that any and every Arab country can, or should, undertake military industrialisation.

By correlating imports, defence budgets, population, and basic economic indicators, a matrix can be formed by which the potentiality of each country for military production can be assessed: does it possess a specified 'critical' combination of factors that allow the creating or expanding of the needed scientific-industrial base or not? Furthermore, although the relevant data are not publicly available, as noted earlier, a breakdown of defence expenditure might indicate other areas in which the development of local manufacturing and technical capabilities may be better exploited. For instance, it is possible that import-substitution of tanks or aircraft will not allow larger or easier savings to be made, but that servicing them locally will. The latter option, and also actual arms production, is especially feasible in light of significant equipment commonality among the Arab armies.

PART TWO

Arab Arms Industries

4

Egypt: The Veteran Arab Arms Producer

Without a doubt Egypt is the foremost arms producer in the Arab region, in terms both of volume and variety. Its military industry has a stated capacity worth \$1.5bn annually, with actual production reaching an estimated \$400m in defence products and roughly the same value of civilian goods in 1987, the latest date for which estimates are available.[1] In 1982, total military exports worth some \$1bn were achieved, later levelling off at \$300–500m annually in 1983–88. Defence production is undertaken by around 30 principal factories and companies, which together employed up to 100,000 persons by the mid-1980s.[2]

Egypt also has the longest history of industrial and military production among the Arab states. Affecting its current situation are its past preponderance in regional politics, its demonstrated ability to develop ties with both superpowers and with the European and the Third World nations, and its large population base. The Egyptian role in the Gulf crisis and war of 1990–91 – winning it new American arms sales, a write-off of \$7bn of military debts owed to the USA, and a potential role in Gulf security – may provide additional funds for the Egyptian arms industry and increased demand for its products. Taken together, these factors suggest that Egypt will remain the leading force in Arab military industry in the foreseeable future, especially following the eclipse of Iraq, which had until 1991 threatened to undermine Egyptian predominance.

The Historical Background

The origins of Egyptian military industry stretch as far back as the 1820s, when then ruler Mohammad Ali first founded a number of factories producing small arms, artillery, ammunition, and warships. A long period of decline followed, lasting until the aftermath of the 1948 Palestine war, when licences were obtained to part-manufacture and assemble a West German trainer jet (1949), a British fighter (1951, subsequently cancelled), and Swedish small arms and ammunition (1952).[3] In charge of this activity initially was the Directorate for Military Factories, which was established in 1949.[4]

Driving Egyptian interest in local production was the perceived need to avoid the shortages of vital arms and ammunition and the dependence on external suppliers that had marked and weakened the military effort during the Palestine war. A further stimulus was provided after the rise to power of the Free Officers and of Abdul-Nasir. The abrogation of the Anglo-Egyptian Treaty of 1936 and later disagreement with the USA soured relations with the West, prompting a turn towards the USSR and its allies for arms. This coincided with a budding regional arms race with Israel. As Mohammad Selim argues, the desire to safeguard national security, achieve weapons self-sufficiency, and stimulate economic development all worked together to suggest to the new, highly nationalistic Egyptian leadership that local military production was the answer.[5] This, it was thought, would break the big power arms monopoly and strengthen non-alignment, while enhancing Egypt's status within the Third World.

Following the decision to expand the arms industry in 1953, several manufacturing plants were set up. By the mid-1960s these totalled five ordnance and two aircraft factories. Their activity was controlled by the Directorate for Military Factories, which was reformed into the National Organisation for Military Production in 1957. Ultimate responsibility for the entire enterprise lay with the Ministry of Defence Production throughout the 1950s and 1960s. During this period, the military industrial sector was marked by its heavy dependence on West European technology. Factories were established by Western companies and handed over on a turnkey basis, while some 600 Western scientists and technicians oversaw development and production in the country. Another distinguishing feature, moreover, was that Egypt sought to develop all the main areas of military production rather than to concentrate on select technologies, for which purpose a number of research bureaux were set up. In this phase, ambitious programmes sought to develop quasi-indigenous aircraft, aircraft engines, and surface-to-surface, long-range missiles.

A consequence of the approach adopted to attain military industrialisation was that Egyptian factories were already suffering from over-capacity by the late 1950s. This was partly a result of lack of co-ordination between the Ministry of Defence Production and the armed forces. On one hand, there was no liaising body between them. On the other, the Western orientation of the industry – its principal directors and scientists were Western-trained, and it relied heavily on Western design input and technology – clashed with the predominantly Soviet leaning of the equipment and doctrine of the armed forces. Not only did Western-type armaments not suit the Egyptian military, therefore, but the defence industry was unable or unwilling to cater to the latter's specific combat needs and custom-build equipment.

Another cause for over-capacity (measured against actual demand for products) was that local products were costly, leading to reduced demand. This was due to dependence on imported technology and machinery and to the need to acquire production skills locally. Faced with such pressures, some three-quarters of military industrial capacity was diverted to civilian production from 1958 onwards. Those plants that still focused on military production continued to incur major financial losses, leading to several programme shut-downs by 1969. In that year the Ministry of Defence Production was also abolished, with responsibility for its activity passing to the Ministry of Industry.

It was at this point that a new potential avenue for the development of local defence production opened up. This was the Soviet offer to allow licensed manufacture of certain military items in Egypt, made in December 1970. Some foundations had already been laid for such a venture. To begin with, the Egyptian armed forces relied almost exclusively on Soviet weaponry, making a decision to produce similar equipment a logical one. In the past, the USSR had been reluctant to extend such permission (except for minor items of ammunition and a few spare parts), and, as Soviet military hardware was so cheap to import (some one-third the price of equivalent items in world markets), the Egyptian armed forces had previously seen little value in producing substitutes locally at greater cost and of lower quality.[6] But during the Egyptian-Israeli War of Attrition, President Nasir obtained initial Soviet agreement to co-operate in the development of the local military industry.

The USSR was already assisting Egypt's civilian heavy industry, so extension into the military domain made sense. In July 1968, for example, Dr Aziz Sidki, Minister of Industry, Petroleum and Mineral Wealth, declared that Egyptian shipyards in Alexandria would construct a number of vessels for the USSR, for which purpose the latter would provide expertise and training (already having provided the original design and machinery for the shipyards).[7] However, Egyptian officials have subsequently accused the USSR of seeking to undermine independent arms production. To support this argument, Minister of State for Defence Production General Jamal as-Sayyed has pointed to the abolition of the Ministry of Defence Production in

1969 and the reallocation of its military factories to the Ministry of Industry, which he sees as proof of a negative attitude to separate, local military industrial activity.[8]

Be that as it may, at the end of 1970 the USSR offered licensed production of certain of its helicopters, radars, and ammunition types.[9] However, this came too late and President Sadat did not respond effectively, although helicopter parts and some light arms and ammunition types were produced. Besides his own political preference for increased dealings with the West, a factor affecting Sadat's decision was Soviet reluctance to allow transfer of technology. Thus the USSR insisted consistently, even at the height of Soviet-Egyptian military co-operation in 1967–70, that all Soviet-designed combat aircraft engines and other major items be returned to the USSR for regular maintenance and overhaul, an attitude that hindered development of local Egyptian skills or the use of existing aviation facilities in Egypt.[10]

A lull lasting several years ensued in the evolution of Egypt's military industry following Nasir's death in September 1970 and the advent to power of Anwar Sadat. This was despite moves by the new Egyptian leadership to revive and reorganize the industry. Motivating the former were strong economic concerns, such as unemployment, foreign debt, and the balance of trade deficit. Increased arms production and exports were seen as ways of improving the situation and generating revenue. But after talks with Britain and France had failed in the early 1970s, Egypt chose a completely new course. This was to establish the Arab Organisation for Industrialisation (AOI), the founding charter of which was signed on 29 April 1975, with several other Arab states. The latter provided the main capital, while Egypt was to offer manpower and infrastructure.[11] Aimed at strengthening Arab security, providing local needs for combat equipment, and exporting surplus production, this enterprise was short-lived, however. The decision by President Sadat to visit Jerusalem in November 1977 and pursue a separate peace with Israel ensured the withdrawal of the other Arab partners.

Nonetheless, the brief experience of the AOI had a major impact. At an immediate level, it reflected a new Arab approach to the issue of military industrialisation and offered a foundation for subsequent efforts in this field. Indeed, with the restoration of Arab-Egyptian relations in the late 1980s, the AOI was suggested once more as an appropriate framework, albeit suitably modified, for inter-Arab co-operation in the field of defence production. At another level, the experience of the mid-1970s later prompted a complete reorganisation of Egyptian military industry, which remains the basis of its current activity. These two aspects are discussed in the following sections.

Joint Arab Military Industrialisation in the Mid-1970s

The shortlived attempt in the mid-1970s to achieve inter-Arab co-operation in defence production is an integral part of the story of the development of the contemporary Egyptian industry, and therefore needs to be reviewed separately.

The first move towards joint Arab military industrialisation came in 1972, when the chiefs of staff of the armies of 18 Arab states met under the auspices of the Arab Defence Council (part of the League of Arab States) to discuss the issue. Their recommendation was to establish a joint project, to which each country would contribute two per cent of its GNP.[12] This proved too ambitious, however, and the issue was only revived in the wake of the October 1973 Arab-Israeli war.

The fourth major round of Arab-Israeli conflict spurred renewed Arab interest in indigenous defence production, for two reasons. The war had demonstrated the unexpectedly high rates of consumption of arms and ammunition in modern combat, and by the same token had underscored the extent to which the main participants were dependent on the superpowers for resupply.[13] Moreover, as General Sayyed observed, experience showed that an army could repair 90 per cent of its hit tanks and 80–90 per cent of all damaged armour *if it had the spare parts to do so*: in 1973 Israel retrieved some 48 per cent of its tanks hit in action.[14] However, the required Arab productive capabilities were either under-developed or non-existent. In contrast, Israel had already deployed a number of locally-designed or locally-produced systems in the war – such as the *Nesher* and the *Kfir* fighter (based on the *Mirage*-III and V) and the *Gabriel* anti-ship guided missile – and immediately sought to expand local production (especially of ammunition of all types and calibres), as well as to stockpile American-made *matériel*. Finally, the dramatic increase in oil prices following the October war magnified the financial resources of the Arab oil-rich states virtually overnight, providing the surplus needed to initiate ambitious, new industrial projects.

The first public indication of Arab interest came with the announcement in March 1974 that the Military Consultation Body of the League of Arab States, also attended by the chiefs of staff, was studying prospects for joint defence production.[15] This was based on a report prepared by seven countries: Iraq, Syria, Egypt, Saudi Arabia, Libya, Algeria, and Morocco. Specifically, the study group recommended the formation of a fund with $1.2bn to finance joint production.[16] Also reflecting the general growing awareness among Arab leaders of the importance of this issue, then Saudi Vice-Premier and Minister of the Interior Prince Fahd stated the following month that his country would insist on receiving 'economic, industrial, technological, and military' projects from the West in payment for oil, rather than money.[17]

In the event, not all Arab states went forward with these plans. Instead, Saudi Arabia, the United Arab Emirates, Qatar, and Egypt came together in mid-1975 to establish the Arab Organisation for Industrialisation (AOI). In effect, the first three countries were to provide little more than funds, while the last was viewed as the main source of infrastructure, technology, and labour. Initial capital of the AOI stood at \$1.04bn – with equal shares of \$260m from each participant – and was supposed to build up to a capital reserve of over \$9bn, though in fact it did not exceed \$1.4bn before the Egyptian-Arab rift in 1979.[18] Administratively, the AOI was answerable to a board of directors comprising the defence ministers of the four member-states.

At the general level, the AOI was intended to achieve several goals: a measure of self-sufficiency in conventional hardware (weapons, ammunition, and spare parts), reduced military production costs, a basis for advanced industrialisation in the Arab region, and a source of income through the export of surplus production to Arab, Islamic, and other Third World countries (especially in Africa).[19] The AOI's founders had few illusions about achieving full self-sufficiency, but felt that by establishing an advanced industrial base the Arabs could supply certain needs, deprive Israel 'and its supporters' of the use of the oil weapon, enable Arab armies to fight longer, achieve inter-Arab co-operation and co-ordination and prevent duality or redundancy of capabilities, and train Arab technical manpower.[20]

The subsequent development of the AOI reveals several features. One was that a basic underlying assumption, at least within the Egyptian leadership, that the AOI was designed to help to turn Egypt into a major producer and exporter of arms. Reflecting this view and the efforts to encourage the process, the Egyptian Minister of Finance, Dr Ahmad Abu Ismail, announced in September 1975 that the Organisation would be exempted from all taxes, fees, and customs duties on its activities and imports, such as for parts, raw materials and intermediate systems.[21] The emphasis on developing Egypt as the main base for the new joint military industry was also evident in the fact that no facilities were established in any of the other partner countries. In fact, it was only as late as the end of 1977 that the AOI director Ashraf Marwan announced that the body would open a branch office in Saudi Arabia, the UAE, and Qatar.[22]

This imbalance in Egypt's favour reflected a divergence of views and goals among the AOI partners, and among the Arab states more generally, that inhibited broader participation initially and that was finally to contribute to the withdrawal of non-Egyptian participation in 1979. Another indication of these differences was the Libyan announcement in June 1978 of the formation of a General Corporation for Defence Production, subordinate to the Armed Forces General Command.[23] By way of contrast, broader Arab co-operation was achieved in non-military industrial projects. An example was the joint company formed at the end of 1978 by Iraq, Kuwait, Jordan, Saudi Arabia, and Syria with a capital of ID150m, to produce cars, tractors, and agricultural machinery (after the Camp David meetings, Egypt was no longer invited).[24]

A second distinctive feature of the AOI was its adoption of an industrialisation strategy based on the acquisition of technology and production capabilities through licensed production from the West. This was to be achieved through assembly or manufacture under licence in either of two frameworks: wholly-Egyptian owned factories, or joint-venture companies formed between the AOI and Western multinationals in which the former would retain majority holdings. Four Egyptian government-owned factories were immediately handed over to the AOI in 1975, while one more Egyptian company and four joint ventures were established over the next four years. During this period the major contracts for licensed assembly or production involved the following hardware: *Lynx* WG-13 helicopters, GEM engines for the preceding, *Alpha Jet* trainer/support aircraft, Larzac and SNECMA M-53 engines for the preceding, electronic equipment, and GM jeeps.

Another area of Arab reliance on Western assistance was in developing trained manpower, who could not only increasingly assume management of the industry and of its main technical processes but also establish the scientific basis for the absorption of advanced military technology and the local infrastructure for an indigenous R&D effort. Towards this end, some 2,500 Arab nationals were sent to the USA and Western Europe to receive education and training, the costs of which effort were included in the AOI budget and capital costs.[25] At the same time, an Arab Institute for Aerospace Technology was established in Cairo with the help of American, British, and French experts, while an Institute of Technical Studies was founded in the Saudi city of Dhahran in accordance with Saudi-British protocols signed in 1973 and 1977.[26]

Most of the contracts signed by the AOI and Western multinationals and governments were negotiated in 1977 and 1978, and were only expected to come into implementation towards the end of the decade. However, President Sadat's policy of concluding a separate peace with Israel brought the whole venture into doubt. At first the response was restrained, as Egypt's Arab partners in the AOI continued to provide funding and approve new contracts throughout 1978. This apparently continued even after the decision by the Baghdad Arab summit held in November to impose a blanket econ-

TABLE 4.1 AOI projects, 1975–79

Type	Quantity	Licensee	Date	Value	Supplier
Lynx WG-13 helicopter	280	ABHCo	1975	£330m	Westland
GEM engine for *Lynx*	730	ABECo	Feb. 1978	$205m	Rolls-Royce
Alpha-Jet aircraft	160	AFACo[a]	Sep. 1978	na	Dassault-Breguet/Dornier
SNECMA Larzac M-53 engine for *AlphaJet*	ha	AFECo[a]	Nov. 1978	na	SNECMA
Swingfire ATGM	thousands	ABDCo	Dec. 1977	$73m	BAe
GM *Jeep*	12,000	AAVCo	1978	$30–35m	American Motors Corp.
Electronics	ha	AECo	1978	na	CSF-Thomson

Source: R. Vayrynen, 'La lente construction de l'industrie militaire egyptienne', *Le Monde Diplomatique*, October 1980.

Notes:

[a] According to the source for this table, these are the Arab–French Aircraft and Engine Company, respectively. However, they do not exist, and it is Factories 36 and 135 that undertake these projects.

omic boycott of Egypt. Thus at the beginning of 1979 and even as the date for signing the Camp David Accords drew near, the new head of AOI, Sheikh Faysal bin Sultan al-Qasimi (Ras al-Khaimah), was still insisting that the Organisation would continue to function unaffected.[27] He moreover invited all member states of the Arab League to join the AOI, an invitation that had also been made nearly a year earlier (after Sadat's visit to Jerusalem) by Qatar's Sheikh Khalifah.[28]

Once Sadat had concluded the Camp David agreement, however, the picture changed radically. Saudi Defence Minister Prince Sultan bin Abdul-Aziz announced the suspension of the AOI and all its enterprises from 1 July 1979 and the annulment of all contracts underway.[29] These were worth some £490m, though the Western companies concerned were able to recoup much of their losses through new arms deals with Saudi Arabia and Qatar. Saudi Arabia moreover decided to withdraw its share of AOI funds and set up a special committee to oversee liquidation.

The latter decision ran into severe difficulties, however. As Shaikh Qasimi explained, of a total of $1.4b allocated to the AOI by then, some $200m had already been spent, much of it on capital investment and machinery, of which Egypt now declared possession.[30] The remaining part of the amount had been invested or placed in various banks. By 1988, the sum of $690m ($662m according to some sources) was still frozen in 18 banks in eight countries, having grown from an original undistributed capital of $260m.[31] Nor was this the only complication. Litigation concerning the suspension of contracts was still continuing at the end of the 1980s, with Westland suing the Egyptian government for £200m.[32]

In 1979, with this course of events, hopes for a joint Arab military industry ended, and it was another decade before Arab and Egyptian leaders could consider reviving the issue. But the brief life of the AOI in its initial form enabled Egypt to develop its own military industry rapidly along new lines.

Organisation of the Egyptian Defence Industry

As indicated earlier, ultimate responsibility for the military industry in Egypt in the 1950s and 1960s lay with the Ministry of Defence Production. The latter exercised its control through the General Corporation for the Military Factories from the mid-1950s onwards. In 1964 this was replaced by the General Establishment for the Military Factories, which was now required by law to make a profit.[33] Not included in the new body was the Egyptian General Corporation for Aviation, which retained responsibility for the aircraft industry. Eventually both were merged in the General Establishment for Military Factories and Aviation, which was placed under the Ministry of Industry in spring 1968, as the Ministry of Defence Production had just been abolished.

The current organisation of the Egyptian military industry differs substantially from that of the period lasting until 1975. At the top political and administrative level, government control is once more nominally channelled through the Ministry of Defence Production, which President Sadat restored in 1971. However, real control rests with the Supreme Committee for Armament (SCA), which consists of the President, Vice-President, Defence, Foreign, and Finance Ministers, and the Minister for Defence Production. In practice, the Defence Minister has traditionally also held the portfolio of Defence Production, though the daily running of the latter ministry is supervised by a Minister of State for Defence Production (General Gamal as-Sayyid in the 1980s). The last also attends meetings of the SCA, as do the Prime Minister, the Minister of the Economy, Armed Forces Chief-of-Staff, and the Chairman and Director-General of the AOI, especially when the annual budget is debated.[34] The SCA is directly responsible for the AOI, and indirectly for the state ordnance factories, which come under the immediate jurisdiction of the Ministry for

Defence Production. A small number of manufacturing workshops run by the armed forces are the responsibility of the Ministry of Defence.

At the practical level, the Egyptian military industry is divided into two main groupings: the Arab Organisation for Industrialisation, and the National Organisation for Defence Production (NODP). The former ran a total of nine production companies at the beginning of the 1990s, while the latter runs 15, grouped into six groups or categories. Both have specialised research and training institutions. As a general guide, aerospace and missile work is concentrated within AOI, along with associated engine, electronics, and avionics plants. Conversely, the NODP is concerned mainly with ordnance, focusing principally on land-based weapons and their ammunition, along with the associated chemical and metallurgical industries. Together, the AOI and NODP are claimed to assure Egypt relative self-sufficiency in small arms and ammunition (up to 90 per cent), and to attain levels of local content reaching 93 per cent in APCs and artillery, allowing for the local production of as much as 60 per cent of the total hardware needs of the armed forces.[35] The specific organisation and activities of the local military industry and of its branches are discussed next.[36]

National Organisation for Defence Production (NODP)[37]

The NODP operates 15 factories, established at various times between 1949 and 1976, which have been run as independent companies since 1964. These are state-run and fully state-owned ordnance factories, with a workforce variously estimated at 50–70,000 in the 1980s.[38] Formerly each factory was identified simply by a number (a multiple of 9 or 10, depending on its geographical location), but now all are known by proper names. The 15 are moreover grouped together according to the nature of its activity and produce.

Despite NODP's long existence, the 1970s represented a nadir in its activity, with annual military production value dropping to a declared $22m by the middle of the decade.[39] Most of the focus in this period was on manufacturing spare parts and ammunition that were reverse-engineered from Soviet designs. The late 1970s witnessed 20 per cent annual growth, however, taking output for the Egyptian market to $270m in 1981. Iraqi demand during the Iran-Iraq war allowed further expansion, though NODP defence-related production has not exceeded the $300m level reached in 1985.[40] Meanwhile, it should be noted that the NODP undertakes manufacturing for the civilian market, with a reported capacity of $115m at the beginning of the 1980s.[41] By 1987, the NODP and the AOI were together allocating some 45–50 per cent of their output to civilian production, according to the then Defence Minister Abdul-Halim Abu Ghazalah.[42]

Arms Production Factories Group

Abu Za'bal Company for Engineering Industries (Factory 100): Most recent of the ordnance factories, this was founded in 1976 to produce anti-aircraft (AA), tube artillery, and tank guns. It covers 500,000 square metres, of which 120,000 are devoted to production facilities, in Abu Za'bal (Qalyubiyya). The Abu Za'bal Company has reverse-engineered Soviet weapons, produced hybrid systems based on the combining of items of non-Egyptian origin according to local designs, and manufactured other weapons under licence. One of the last items is the British L-7 105mm tank gun, produced with technology and expertise from Royal Ordnance.

Ma'adi Company for Engineering Industries (Factory 54): Founded near Cairo in 1949, this factory started production of Swedish-designed small arms in 1954. It has since concentrated on small arms, reverse-engineering the Soviet AK-47 (*Kalashnikov*) and producing locally-designed pistols.

Helwan Company for Machine Tools (Factory 999): In addition to sighting equipment and spare parts, this company produces mortars of 60/82/120-mm calibres and 122-mm rocket-launching tubes.

Ammunition Production Factories Group

Helwan Company for Engineering Industries (Factory 99): Specialises in manufacturing metal parts and casings for medium and heavy ammunition. Thus it has also developed expertise in developing and manufacturing anti-armour warheads and projectiles, runaway bombs, and artillery rockets.

Heliopolis Company for Chemical Industries (Factory 81): Produces a wide variety of complete rounds for medium and heavy weapons, including AA gun ammunition of 57mm and above, 122-mm and 130-mm tube artillery, 105-mm tank shells, mortar shells, and unguided artillery rockets. It also manufactures mines, illuminating and smoke rounds, some naval munitions and air bombs, warheads for man-portable anti-tank and anti-aircraft guided weapons, and sapper charges. The company comprises two main factories and seven minor ones: the first two manufacture the above-mentioned products, while the rest specialise in emptying used ammunition casings, preparing napalm and flamethrowers, manufacturing fuses, and producing explosives, necessary chemicals, and rubber- or plastic-based items, such as gas masks and tank wheels. During the first half of the 1960s, this company included the Heliopolis Missile Plant, which developed the long-range, heavy, surface-to-surface bombardment rockets known as *al-Qahir*, *al-Zafir*, and *al-Ra'id*.

Shubra Company for Engineering Industries (Factory 27): Founded in 1953, this company has been producing small arms ammunition since 1954. It switched from Western designs to Soviet models in 1956.

Abukir Company for Engineering Industries (Factory 10): With 1,630 employees, this company produces standard calibre ammunition for Western and Eastern assault rifles, as well as varied items such as smoke grenades, mortar cartridges, and individual rifle-cleaning equipment. It was founded near Alexandria in 1952 and commenced production in 1954.

Al-Ma'asrah Company for Engineering Industries (Factory 45): Has specialised since its foundation in 1956 in producing ammunition for medium-calibre AA guns, anti-personnel mines, hand grenades, and fuses for heavy ammunition.

Chemicals and Explosives Production Factories Group

Abu Za'bal Company for Specialised Chemicals (Factory 18): This factory commenced production in 1957, seven years after its foundation, and now has 4,000 employees. It produces a complete range of propellants and explosive charges for small, medium, and heavy calibre land weapons, as well as for mines, bombs, rockets, and the *Swingfire* ATGM.

Qaha Company for Chemical Industries (Factory 270): Its 1,500 employees concentrate on the manufacture of training, tracer, and smoke rounds and components, and on assorted items, such as hand grenades and military batteries and chargers.

Metallic Production Factories Group

Helwan Company for Non-Ferrous Industries (Factory 63): From providing metallic materials for small-arms ammunition factories in 1954, this company has expanded to producing the alloys needed for the manufacture of cartridges, bullets, and certain heavy-calibre components. Its 4,000 employees also use a variety of processes to produce strips and bars and semi-finished components, such as sand- and die-casting, rolling, and extrusion.

Helwan Company for Ferrous Products (Factory 9): Provides iron castings for all other military factories, as well as engines, machine tools, electrical motor bodies, and associated components.

Helwan Company for Metallic Equipment (Factory 360): Since 1964, its 2,175 employees have produced sheet metal, which they use to make mines, sapper charges, and for incorporation into trenches and firing positions.

Electronics

Benha Company for Electronic Industries (Factory 144): This is one of the most advanced ordnance factories, with some 380 graduate engineers among its 3,000 employees (having grown from 150 employees in late 1981), some 70 per cent of whom are either engineers or skilled technicians.[43] Its products include a variety of receivers and transceivers for military radios and telephones, jammers, mine detectors, and timers and explosive devices. Sales were reported at E £70m in 1986; alternatively annual turnover was estimated at $45.6m in 1987 (E £100m).[44] Britain's Plessey is involved in producing items such as tank and squad radios – 2,500 PTR-1410E tank radios are produced yearly using one shift – while American Westinghouse oversees the licensed production of the AN/TPS-63 radar.

Field Service Equipment

Helwan Company for Diesel Engines (Factory 909): From producing engines for T-34 tanks in 1964, this company has moved to manufacturing a range of engines from 8 to 150 hp.

Finally, the NODP now runs two additional, major production facilities. These are the *Abu Za'bal Tank Repair Factory* and *Factory 200*; the former is responsible for overhauls on American-designed M-60 tanks, while the second, construction of which was completed in 1991 is undertaking production of the new M-1.

Arab Organisation for Industrialisation (AOI)

Following the Arab decision to withdraw from the AOI, it was wholly taken over by the Egyptian government, which assigned Ahmad Zando as its new head, in accordance with statutes promulgated by Parliament formalising the Organisation as an Egyptian institution.[45] A new board was appointed to run the AOI, including among its members the Deputy Governor of the Central Bank, the Assistant Minister of Defence, Secretary of the Finance Ministry, and the Vice-President of the Republic.[46] At the time it had only $40m left as starting capital, but $200m's worth of new equipment and infrastructure.[47] Most major production projects had to be stopped, though several minor ones continued in co-operation with Western partners.

By 1990 the AOI had 20,000 employees, working in nine companies.[48] Of these, five divisions are state-owned in their entirety. (Four of the five were formerly military ordnance factories, while the fifth is a later venture.) The remaining four subsidiaries are joint-venture companies between the AOI and Western companies and multinationals, though contracts have since been issued to set up several more. In theory, the latter are also open to private Egyptian capital, but total government control, the secretiveness of the sector, and limited investment opportunities for local industrialists and businessmen have effectively deterred such participation until recently.[49] The AOI now includes all aerospace facilities and most plants involved in rocket and missile work. Since October 1987 it has been headed by General Ibrahim al-'Orabi.

$300m's worth of output was offered in 1985, but this seems exaggerated. Following his appointment as head of the Organisation in late 1987, General al-'Orabi stated that he had found little industrial activity there. He later estimated military production by NODP at $152m in 1987, most of which was supplied to the Egyptian Army.[50] The SIPRI figure for that year is even lower, at $100m. Yet the higher production figure of $300m may nonetheless be explained in part by the fact that the AOI is actively involved in civilian industry as well, in which case undifferentiated output may refer to both military and civilian goods. Among the Organization's non-military products are water-desalination devices, solar heaters, and plastic wares and furniture. It is also involved in joint ventures with other Egyptian firms, such as the Egyptian Corporation for Canning and Packing to manufacture spare parts for machinery used in packing, transporting, and extracting oils.[51] Furthermore, the AOI took a conscious decision at the beginning of 1990 to develop its civilian lines and diversify their products, while the government moved to sell off public sector assets to private industry.[52] An example of further expansion is the joint venture signed in 1990 with the British company International Dynamics, for a $5m plant to refurbish machine tools. AOI companies are as follows:

Saqr Factory for Developed Industries (Factory 333): Mainly produces artillery rockets and light guided missiles that are reverse-engineered and improved versions of Soviet weapons. Other products include anti-tank grenades, illumination and smoke-laying rockets, and targets. The factory, which covers an area of 2.3 square kilometres in Heliopolis and includes two test ranges, was originally established in 1963 to produce Soviet-designed, surface-to-surface guided missiles and spare parts for aircraft. By 1982 it had a workforce of 5,000–5,500, of whom 400 were graduate engineers, and boasted a high-quality plating and metal finishing facility which uses European-forged alloys.[53]

Aircraft Factory (Factory 36): This facility is responsible for most of Egypt's ambitious aircraft assembly and production plans. It already assembles the *Alpha Jet* and the Chinese F-7 and would do similar work on the *Mirage*-2000 Northrop F-20, or F-16 *Falcon* combat aircraft if the relevant deals were struck. Factory 36 also undertakes MiG-21 rebuilds and Su-7/20 overhauls, as well as making parts for both aircraft.[54] It employs 3,000–3,500, who include 1,500 production workers and 150 engineers, besides technicians and managerial staff. Founded in Helwan in the late 1950s, it originally undertook development and production of the HA-200 trainer and prototype HA-300 fighter, and still retains the surface treatment plant established at that time. Factory 36 moreover works in coordination with a subsidiary facility at Tanta that maintains *Mirage*-2000 aircraft and assembles its *Magic* AAM.

Qader Factory (Factory 72): Responsible for the production of the *Gumhuriyyah* trainer plane from West German designs in the 1950s, this factory now assembles the Brazilian *Tucano* trainer under licence. It also produces German- and locally-designed APCs, armour, and air bombs.

Helwan Engine Factory (Factory 135): Here 3,500–3,800 employees (including 1,400 production workers) produce engines – part-assembled and part-manufactured – for the *Alpha Jet*, as well as spare parts and undertake overhauls for other engines. Founded in 1960, this factory also manufactured engines for the locally-produced HA-200 and HA-300 aircraft.

Electronics Factory: Largely an avionics facility (founded in 1978), this is the main centre for producing and developing land communications equipment, aircraft communications and radar, and avionics and electronic circuits. Some of the last are used in licence-assembled aircraft or exported back to the licensing mutlinationals.

Arab-American Vehicle Company: First of the joint-venture companies, it was formed in 1977 between the AOI (with 51 per cent) and American Motors (49 per cent). Allocated 70,000 square metres, it was designed to produce 12,000 military jeeps, with a capacity to produce 10–12,000 of them annually.[55]

Arab-British Helicopter Company: This company was formed between the AOI and British Westland, following lengthy negotiations that began in 1974 between France, Britain, Saudi Arabia, and Egypt, over the assembly under licence of *Lynx* helicopters in Egypt. The project collapsed in 1979, however, resulting in litigation that is still unresolved. ABHCo now produces the French *Gazelle* – at a maximum rate of four per month – and parts for the *Puma*. Its factory covers 16,000 square metres, divided into two main assembly shops of 3,600 square metres each, and many of its technicians have received their training abroad.[56]

Arab-British Engine Company: Established at the same time as the preceding one, this company involved the AOI (70 per cent) and Rolls-Royce (30 per cent), with a starting capital of \$20m.[57] It immediately signed contracts with Rolls-Royce worth £100 million. Allocated 50,000 square metres, its 3,500 employees (some sources state only 400 employees)[58] produce engines for the *Gazelle* and parts for other helicopters. They also manufacture parts and undertake overhauls for Mi-8 engines through reverse engineering, and are claimed to run the best Tu-2 engine-test facility outside the Soviet bloc.

Arab-British Dynamics Company: Formed between the AOI and British Aerospace Dynamics, this joint venture produces guided missiles, based on Western and Soviet designs.

Finally, in mid-1989 the AOI announced that it intended to establish a new electronics factory, and invited bids from local and foreign companies. A deal was subsequently signed with a French group to complete setting up an electronics firm by the end of 1991.[59] Establishment cost £E60m (\$23.3m), and the French were to receive FF400m (\$69m) to operate and maintain the factory until 1994. Products were aimed at both the civilian and military markets, and included television and video sets. Other plans envisaged producing a light civilian aircraft and setting up lines to make industrial boilers and heat exchangers, while AOI also multiplied its contracts to assemble foreign-designed cars and buses.[60] It also concluded a deal for the production of aluminium panels for the F-16C/D, worth a total of \$500,000, following an American-Egyptian agreement that Egypt could contribute to the fourth 40-plane batch it proposed to buy of the fighter.[61]

Army Factories

In addition to the state-owned ordnance factories and state-controlled joint ventures, the Egyptian armed forces also run a number of manufacturing and service plants. Among the latter is the

Air Force's main aircraft overhaul depot at Helwan, which employed 3,900 personnel (including 920 civilians) at the beginning of the 1980s, to undertake work on Soviet-designed aircraft.[62] The EAF also runs the facility at Tanta for *Mirage*-2000 maintenance and *Magic* AAM assembly, and separate plants for maintenance of the *Hawk*, *Crotale*, and *Chaparral* SAMs (and supposedly, modification of the latter two), and planned to establish a joint repair and assembly plant for the *Sidewinder* AIM-9P-3 AAM.[63]

Army workshops have also undertaken modifications and modernisation on T-54/55 and T-62 tanks, while small plants have produced uniforms and similar non-combat gear. Similarly, *October* (copies of the Soviet *Komar* class) FAC/Ms have been produced at the Central Naval Centre west of Alexandria.[64] Finally, the armed forces also undertake construction projects, such as the 'military cities' being built near Cairo and Alexandria.

Private Sector Companies

Although the incentives have been limited, as described previously, a small number of private companies have entered the defence production market in Egypt.

Arab International Optronics (Law 43 company): The largest and most dynamic private company, this is in fact a joint venture between an Egyptian Defence Ministry subsidiary (51 per cent) – the National Service Project Organisation – and a British company (49 per cent), United Scientific Holdings.[65] The latter owns a number of British and American subsidiaries, among which is Alvis, the company producing the *Scorpion* light tank that Egypt had shown interest in producing under licence in 1981.[66] Arab International Optronics was established in May 1982, after the Army had spent $18m on sights and other optical equipment, and claims to have started production in January 1984. Its 280–300 employees produce laser range-finders, night-vision devices and similar equipment (such as binoculars) at a factory in Cairo's suburbs covering 9,700 square metres. AIO's operation was based on a five-year Army requirement with guaranteed sales for the first two years.[67] Starting with an initial capital of E £4.5m in 1984, its investment value multiplied nine-fold in the first three years of its operation, and it was achieving a E £3m profit by 1986.[68] By autumn 1988 the company head, General (Eng.) 'Adel Qinnawi, could boast that it was producing 80 per cent of import needs of military and civilian electro-optical goods.[69]

Compuland Company: Specialises in developing computer software and in training personnel.

High Tech: This company boasts links with 20 major manufacturing firms to produce wiring systems, using imported designs, connectors, and electronic components, as well as spare parts for military and civilian aircraft.[70]

Electrolab: Supplies radios, antennae and similar equipment under licence to the armed forces from a factory in a Cairo suburb. Not yet involved in exports.

Egyptian Specialised Electronic and Training Equipment Company: Provides electronic firing-ranges designed to save ammunition and money in training.

Finally, a number of new joint ventures involving Western companies or multinationals and Egyptian firms were being established at the end of the 1980s.[71] One involved a military vehicles plant, with the participation of West Germany's Daimler-Benz, and Misr Car and government banks on the Egyptian side. Spain's ENASA was also supposed to found a military truck plant, although Egypt's subsequent financial problems led to the suspension of the plans, while American General Motors and other companies are to establish new defence-related production lines at their automotive and electronics factories in Egypt.

Military Products

As mentioned earlier, Egypt's military industry produces goods for a variety of uses. The principal categories are discussed below.

Aircraft

The aerospace industry is probably the most developed branch within the Egyptian military industry. Yet much of its activity is confined to the assembling under licence of foreign aircraft. With one exception – the Chinese F-6 fighter – all locally-produced aircraft (including helicopters and drones) are Western-designed. In addition to whole systems, Egyptian companies manufacture components and spare parts for some aircraft and for certain engines, often for re-export and use in aircraft being sold to Egypt. This is again generally done under licence, except in the case of Soviet-designed aircraft and helicopters. In the latter case some parts and engine components were reverse-engineered after the virtual breakdown of relations with the USSR in the 1970s.

The one programme in which an appreciable level of local production has replaced assembly from kits is the *Tucano* trainer, followed at a lower level by the *Alpha Jet* trainer/ground attack aircraft. An example of the limitations of local aerospace industry is the request, in mid-1985, that the Swedish Swedair Company convert four DHC-5 *Buffalo* STOL transports into flying classrooms – by installing 10 operator seats with full avionics equipment,[72] indicating the lack of flexibility in local production lines (tooled for specific projects only) and insufficiency of electronics centres.

A summary of the various aircraft programmes contemplated or underway in Egypt illustrates the preceding limitations. The aviation industry started with the part-manufacture and part-assembly of a World War Two-vintage German trainer, and then sought to produce the British *Vampire* fighter under licence, in 1949–51. The only real attempts at indigenous design were the HA-200 trainer and HA-300 fighter in the 1960s, both of which were, however, based on Spanish designs originally. Moreover, Western scientists formed the backbone of research and development teams working on these projects. An extended hiatus followed, broken in the second half of the 1970s with a series of new contracts. Since then, the Egyptian aircraft industry has assembled the Brazilian *Tucano* trainer, the French-British *Alpha Jet* advanced trainer/support aircraft, the Chinese F-7 fighter, and the French *Puma* and *Gazelle* helicopters. It has also sought to assemble and then part-produce the British *Lynx* helicopter, and the American F-16 and the French *Mirage*-2000 fighter.

As in other cases in the Third World, the purpose behind this industrial strategy has been to acquire the necessary production skills and military technology gradually, with the eventual aim of producing indigenous systems. Expressing this view, the former AOI head Ahmad Zando declared late in 1985 that, 'the highest goal is to design and produce an aircraft that is 100 per cent Egyptian'. According to this approach, complete weapon systems are at first produced from imported sub-assemblies, after which comes full assembly based on a majority of imported parts and a few locally-manufactured ones, steadily increasing the latter's proportion until total indigenous production can be attained. Some headway has been made in this direction in Egypt, which claims both to be producing (rather than assembling) 73 per cent of the *Tucano* trainer and to be capable of reaching a level of 100 per cent. However, the accumulation of expertise and development of skilled manpower has been slower than expected, leading both to time delays and to lower levels of local input than were originally planned for each stage.

Additional problems have dogged Egypt's efforts to develop an advanced aircraft industry. A major one is securing finance. According to General Mustafa, head of the Egyptian Armament Authority, Egypt can rely on itself for up to 60 per cent of its needs, all that is lacking is money.[73] Lack of funds has caused delay or even cancellation of several programmes. Most notable were the *Lynx* helicopter and the *Hawk* trainer, originally due to enter production in the late 1970s but halted owing to the collapse of the AOI in 1979, and the *Mirage*-2000, which was meant to go into production in late 1985. More recently, in June 1986 the Egyptian Ministry of Defence Production had to forgo a credit line worth £70m ($107m at current prices) that had been approved by the British government and commercial banks in 1985, owing to Egyptian uncertainty about means of refinancing. An element of the financial problem is the need to have extended production runs in order to achieve economies of scale: the minimum break-even requirement has been estimated at 120 aircraft, but unless the defence budget rises or export orders are guaranteed such levels cannot be attained.[74]

A second problem has been the reluctance of Western companies to transfer technology. Thus the technicians in charge of crucial areas of production, especially including quality control, are often Western. This is especially the case if the manufactured goods – such as parts and component sub-assemblies – are to be re-exported to the original licenser, according to offset arrangements or other contractual obligations supposed to embody the gradual transfer of production technology. Western reluctance is evident, for example, in the difference between the French and the Brazilian approach to production licences: whereas the latter has permitted local manufacture of parts up to a (largely theoretical) total of 100 per cent – of which Egypt claims to have reached 73 per cent – the former set a 15 per cent limit on the share of local manufacture in the *Alpha Jet* programme.

With the starting problems of Egyptian industry and its limited ability (in terms of manpower, infrastructure, and organisation or administration) to absorb new technology, such supplier policies only slow its development even further and delay the planned, staged shifts from assembly into indigenous production according to original timetables. However, it must be said that the main obstacle to greater technology transfer is often the lack of Egyptian funds. According to a leading industry official, were the surplus capital or export orders available to allow longer production runs, the proportion of local manufacture could be raised constantly.[75]

Finally, a third problem is that Egypt has not been completely free in choosing aircraft types to produce locally. In the past this related to Soviet reluctance to permit licensed production of its weapons in Egypt (or in the Third World generally, for that matter). The politically-motivated withdrawal of Arab participation in the AOI – depriving it both of capital and of markets – led to the abandonment of plans to produce the Westland *Lynx* helicopter and the Hawker Siddeley *Hawk* trainer/support aircraft. More recently similar problems have affected American-Egyptian ties. Whether for commercial reasons (fear of competition) or political and strategic ones (altering the regional balance of power or the Egyptian-Israeli balance), the USA in particular has been slow to grant licences. Thus in 1985 it offered Egypt the *choice* between co-production of the export-designed F-20 fighter *or* assembly of the F-16. These problems – financial, technological, commercial, and political – continue to limit Egyptian options and prospects as local industry seeks to achieve ever-more ambitious manufacturing projects.

The following sections deal with the several programmes of the Egyptian aerospace industry.

Trainer Aircraft

• The *Gumhuriyyah*: Egypt's first venture into the realm of aircraft industry was with the Bucker Bu-181D turbo-prop trainer. This was a German design developed during World War Two, for which Egypt obtained a production licence in 1949. Beginning in 1950 the local model was part-assembled, part-manufactured at Heliopolis Air Works (Factory 72). Initially designated the HA-100, it was soon renamed the *Gumhuriyyah*. The trainer was repeatedly updated, with six versions eventually being produced. Some 300 trainers of this type were produced, with production ceasing in the late 1960s. Sales were also made to Jordan, Algeria, Saudi Arabia, and Sudan. And although it has largely been replaced, up to 200 *Gumhuriyyah* basic trainers are still in service with the Egyptian Air Force, nearly 40 years after the first one was built locally.[76]

• The HA-200 *al-Qahirah*: the Egyptian aerospace industry took a major step forward in the early 1960s with the initiation of two ambitious programmes. The first was the production, under licence, of the Spanish Hispano HA-200 *Saeta* jet trainer under the name of *al-Qahirah*. This was assembled at the Helwan Air Works (Factory 36), which was inaugurated in July 1962. Some 100 Spanish workers and technicians assisted in the establishment of this programme, which involved local production of the fuselage and the engine. The latter, the E-201 Turbomeca *Marbore*, was produced by Factory 135 (established in 1960) until 1970. In all, 63 of the tandem, two-seat trainers were produced, of which 30 remain at the factory (12 are in flight status), along with 200 engines.[77]

• *Alpha Jet*: a 12-year interruption followed the cancellation of the HA-200 programme. Finally, Egypt concluded an agreement with France in January 1981 for the purchase and local assembly of the Dassault-Breguet/Dornier *Alpha Jet*. This was to be the first concrete step in the revival of the Egyptian aerospace industry, after the intervening failure of the Arab-Egyptian AOI in 1975–79.

Under the terms of the new agreement, France would supply four *Alpha Jets* in flyaway condition in the MS-1 advanced trainer version in 1982, and four more in the MS-2 ground attack version in 1983. Starting in mid-1982, the AOI's Helwan Aircraft Factory (Factory 36) would commence assembly of another 26 MS-1s and 11 MS-2s, for a total of 45 from all sources. The contract also called for the manufacture of certain parts and sub-assemblies of the aircraft at Helwan, and of the *Larzac* 04-C5 engine, 80 of which were also to be produced under licence by Factory 135. Half the necessary tooling was to be manufactured in Egypt and half by Dassault-Breguet and Turbomeca.[78]

Engine parts to be made locally included high-pressure compressors and turbine parts. Among the parts to be produced locally for the aircraft were the ailerons, rudders, tail cones, flaps, engine bypass ducts, some racking, and a growing proportion of the airframe (mainly the rear fuselage). In late 1983 local production also started of 310-litre drop tanks. The Factory's Avionics Division would also produce, in part or in whole, such systems as the Thomson-CSF HUD, TRT radio altimeter, and V/UHF radio.[79] As it produced them under French quality control, the AOI was able to re-export parts for use on Dassault-Breguet production lines, within the agreed five per cent limit (of total production cost).[80] At another level, in order to achieve commonality with the *Mirage*-5Es in service with Egypt (and the *Mirage*-2000s to arrive subsequently) and so ease support, the MS-2 *Alpha Jets* were equipped with the same attack systems (of French manufacture).[81]

The AOI commenced tooling-up for production by the beginning of 1982, and was delivering aircraft and engine parts within a few months, allowing the first locally-assembled *Alpha Jet* to be rolled out in September (and delivered to the Air Force in November). With aircraft being produced at a rate of 1.5 monthly, the original order was completed in 1985, during which time the share of local input had risen to 48 per cent.[82] Despite the success of this programme, Egypt has faced some difficulty in expanding it, despite holding an option for the production of an additional 45 *Alpha Jets*. The main problem has been capital, specifically through guaranteed exports. Thus decisions by Oman to buy British *Hawk* trainers (instead of *Alpha Jets*) and by Iraq to buy *Alpha Jets* direct from France (or even to produce them locally), have so far closed Egyptian export options.

• *Tucano*: seeking a new primary trainer for its air force, and in order to fulfil a similar Iraqi requirement, Egypt concluded an agreement in December 1983 with the Brazilian company Embraer to acquire its two-seat turboprop EMB-312 *Tucano*. According to the terms agreed, Egypt would buy 12 ready-made models, and then assemble a further 108 aircraft (for a total of 120) using kits provided by Embraer. Also included in the deal was the licensed local assembly of the Pratt & Witney (P&WC) PT6-25C engine to power the *Tucano*. Total contract value was stated to be $181 m.

In addition to the agreed production run, the AOI retained an option to assemble 60 additional aircraft. It moreover signed a non-exclusive sales agreement with Embraer enabling it to make additional sales. Of the total number, 80 *Tucanos* would be exported to Iraq, at a unit price of $1.2 m, with the remaining 40 going to the EAF.

Assured of customers, the AOI moved quickly from the licensing stage in 1983 to production in 1984. Undertaking assembly of the *Tucano* in Egypt was the Qader Factory (Factory 72), while the engine was to be produced at the Engine Factory (Factory 135). The first locally-assembled *Tucano* was flown at an EAF parade in July 1985, having been delivered the month before. This was one of a total of 12 such aircraft produced that year. These were assembled from complete sub-assemblies provided by Embraer, but the AOI's Aircraft Factory started to manufacture a growing number of parts and components for the 48 additional trainers produced during 1986. Indeed, by 1988 Egyptian officials were claiming that 72–73 per cent of the *Tucano* was being locally-manufactured rather than assembled. At the same time, the domestic contribution to the P&WC PT6-25C engine for the *Tucano* reached 40 per cent.

Producing the *Tucanos* at the rate of four per month, the AOI was able to complete the Iraqi order in April 1988. This was also the occasion for the delivery of the first batch of trainers to the EAF, which had taken its full consignment by early 1989. At this point, the AOI decided to take up its original option to produce locally 60 more *Tucanos*, hoping to attract new buyers in the Gulf and Iraq. In the meantime Iraq was itself negotiating with Brazil over possible licence-production of the *Tucano*, as part of the former's efforts to establish a local aerospace industry.[83]

Combat Aircraft

• The *Vampire*: Soon after production of the *Gumhuriyyah* trainer commenced in 1950, the Egyptian government concluded an agreement with the British company De Havilland for the local production of its *Vampire* jet fighter. Before practical steps could be taken, however, the British government cancelled the agreement, following the abrogation of the 1936 Anglo-Egyptian Treaty by Nasir. This false start encouraged an effort to develop an indigenous fighter towards the end of the decade, based in part on Western technology.

• The HA-300: This new effort took the form of an ambitious lightweight fighter programme. The HA-300 was in fact another Hispano design (originating with Willi Messerschmidt), abandoned by Spain at the beginning of the 1960s. In Egypt, further development was supervised by German scientists under Professor Brandner, and several significant modifications were made, such as the inclusion of horizontal tail surfaces and the alteration of the engine air intakes. At the same time, Factory 135 worked to design an entirely new engine, which was to replace the British Bristol Siddeley *Orpheus* BOr.2 turbojet used in the original version (several examples of the engine were sent to Egypt).

One of the most significant aspects of this effort, which was undertaken partly in collaboration with India to equip its own indigenous HF-24 *Marut* fighter, was that it is virtually the only attempt by a Third World country to develop an aircraft engine locally. (Given that the E-300 derived much from the *Orpheus*.) This only underlines the fact that in the 1950s and 1960s, the state of the art in military technology was such that the gap between LDCs and AICs was not insurmountable, in sharp contrast with the situation today.

In any case, 18 prototype E-300 engines were produced for testing purposes. They provided 4,800kg thrust (10,580lb), and were expected to permit a maximum theoretical speed of Mach 2.2.[84] In fact, only subsonic speeds were achieved by the first prototype in its maiden flight in March 1964. The other two prototypes, the only others ever built, were used to test the engine on the ground. Constant difficulties with the engine finally led to the complete abandonment of the programme in 1969.[85] So ended all Egyptian attempts to develop an indigenous fighter.

- The hypothetical choice between F-5, F-20, or F-16: as in the case of local production of trainer aircraft, a long interruption followed the cancellation of the HA-300 project. Leaving aside the ground attack version of the *Alpha Jet* already in local assembly, Egypt's first attempt to revive its fighter programme came in 1984. At this point, interest was divided between the French *Mirage*-2000 and an alternative American-made aircraft.

In the event Egypt made early decisions to purchase the *Mirage*-2000 and the F-16, but the debate continued over possible production of an additional combat aircraft, with an eye especially to export potential. This motivation had, in fact, already prompted the AOI into considering co-production of the F-5 with Northrop Corporation as early as 1981/1982.[86] At that time, the company offered assembly of the F-5E *Tiger* II from knocked-down parts, to be followed by assembly of the F-5G and its F404 engine from kits and eventually by co-production of up to 60 per cent of the airframe and engine (as well as final assembly) as part of a $1.1bn programme including 120 aircraft and 10-year supply of spare parts.[87] Due to envisaged delays in setting up this project and to the fact that the EAF would be fully involved in absorbing its new F-16s and *Mirage*-2000s from 1984 onwards, the F-5 option was dropped. By 1984 the focus was on the F-20 *Tigershark* version of the F-5 which had been developed specifically as an export fighter. Supporting this option, the then Defence Minister, Abu Ghazalah, saw the possibility of selling the F-20 to the Gulf states, and added that Egypt could achieve an 80 per cent local production rate.[88]

Nothing evolved from this debate, however. Nor was co-production with Turkey taken up seriously; this was a possible option under the Turkish-Egyptian military agreement on co-operation, signed in August 1984, at which time Turkey was already setting up its own factory for the production of 160 F-16s.[89] Instead, the Egyptian Air Force opted for purchase of 40 F-16 fighters on a flyaway basis. This was followed by an order for 40 more, to be delivered by 1992. Finally, late in 1985 Egypt was granted permission to establish two lines for the production of parts for the F-16 and its engine, though no actual production was started.[90] And although it had no input whatsoever in producing the first 80 aircraft, the Egyptian aerospace industry still hoped to win a share in the production of a third, and possibly final batch of F-16s.

The latter option had not been approved by the end of the 1980s, and in fact seemed to be virtually written-off by the terms of the preliminary agreement concerning the third 40-aircraft batch of F-16s. Instead, in May 1989 visiting American officials approved the Egyptian request to produce parts for Northrop on a sub-contract basis. According to the deal worth $500,000, the AOI would manufacture 3,000 parts – mainly aluminium panels for aircraft covers, fairings, support structures, and bulkhead parts – to be used in 400 aircraft earmarked for the USAF and other foreign forces.[91] Egypt has thus been unable to move to the level of local production it prefers, despite interest in co-producing a further batch of 40 aircraft, partly due to American reluctance and partly to lack of local funds. When Egypt finally won American approval to buy a fourth batch of F-16s in 1991, these were to come from the Turkish production plants.

• *Xian* F-7: the little that is known about this programme is that Egypt assembled a number (probably several dozen) of the Chinese-designed *Xian* F-7, itself a modified copy of the Soviet MiG-21. The basis for this venture was Egyptian familiarity with both the MiG-21 and Chinese aircraft. During the second half of the 1970s, the Egyptian Air Force cannibalised some 100 MiG-21s in order to save another 100.[92] In 1979 it also purchased 60 Chinese F-6s, copies of the Soviet MiG-19.[93] These came with a number of engines for updating other Soviet-designed aircraft in the Egyptian inventory. However, although Egypt is known to have 52 F-7s in service, these may have been bought directly from China, as the locally-assembled versions were probably sold immediately to Iraq.[94]

• *Mirage*-2000: while the debate continued in Egypt in early 1984 about the pros and cons of acquiring production rights for the F-5/20/16, a decision was taken to start updating the Air Force with the French *Mirage*-2000. This had been under consideration since 1982, when the AOI hoped to win a 10 per cent stake in any potential contract.[95] (At that time, AOI had won an agreement to manufacture small sheet-metal sub-assemblies for the Dassault-Breguet *Falcon* 50 light executive transport aircraft.[96]) The *Mirage*-2000 contract formed part of a wider deal with France that also involved purchase of the *Crotale* SAM and the *Gazelle* helicopter. The *Mirage*-2000 component was worth $1bn. This was for 20 aircraft to be delivered off-the-shelf in 1984, with an Egyptian option to buy 20 more, and a final target of 60.[97]

The additional importance of the agreement, for Egypt, lay in two provisos. The first was that the AOI would set up production lines for the manufacture of a number of parts for both the aircraft and its Snecma M53 engine. AOI would thus act as a sub-contractor to Dassault, which was the primary contractor to the Egyptian Air Force, and would therefore continue producing parts for *Mirage*-2000s destined for other of France's customers. The second was that Egypt would be able to commence local assembly of the fighter, starting from the 25th aircraft – in other words, once delivery of the second batch of 20 had begun.[98] The AOI moreover gained permission to market its assembled *Mirage*-2000s in the Middle East and Africa.[99]

Only the first side of the local production protocol has, in fact, been taken up. Egypt makes some fuselage parts and auxiliary fuel tanks for the aircraft itself, and titanium compressor blades for the Snecma engine (which is assembled at Factory 135).[100] Owing to financial constraints, the Egyptian Air Force has been unable to order the second batch of 20 *Mirage*-2000s, thus depriving the aerospace industry of the opportunity of assembling them. This has not halted efforts at developing local capabilities, though. Talks held between military teams from both sides in Cairo in late 1988 were aimed at producing a new protocol for co-operation that would entail, among other things, the construction of a maintenance centre for French aircraft in Egyptian service.[101] However, major obstacles remain in the way of assembling the *Mirage*-2000 in Egypt, not least of which are the lack of funds and Egyptian interest in acquiring and co-producing the F-16. Another problem is the difficulty of finding export customers; the UAE ordered the aircraft direct from France in 1984. A new opportunity was offered in 1989 by Iraqi interest in buying the *Mirage*-2000, in which Egyptian industry might have been involved, but such participation was precluded by the Gulf crisis of 1990–91.[102]

Helicopters

- *Lynx*: licensed production of the British Westland *Lynx* WG-13 multi-purpose helicopter was the first industrial project planned by the newly-established AOI in 1975. It was also scheduled to be the first new production programme of any type undertaken by the Egyptian aerospace industry since the 1960s, and was therefore supposed to provide the basis for a range of subsequent activities. The explicit aim of the AOI was to standardise requirements among its members and select a single design that would then allow larger and more economical production runs. However, the latter would be surplus to the needs of AOI member-states, in order to provide for exports, which were seen as a major function of the new organisation.

In 1975, the Arab British Helicopter Company was formed in Egypt within the AOI, and signed a deal in principle with Westland for the local assembly of the *Lynx*. A production contract was then concluded in 1978, worth £330m, for a run of 220–280 helicopters. Of these, 20 would be delivered by Westland, starting in January 1980, after which at least 200 more would be first assembled, and then part-produced in Egypt, starting in June 1980.[103] Components would be supplied by Westland initially, followed by those parts requiring special tooling. Total programme duration was set at 92 months from April 1978, at a production rate of four helicopters monthly. The first 50 *Lynxes*, produced in four versions, were earmarked for the four AOI member-states. In February 1978, a second British-Egyptian deal was concluded for the local production of the GEM engine for the *Lynx*. This was to be undertaken by the newly-formed Arab British Engine Company, which signed a contract with Rolls-Royce worth $205m for the assembly of 730 engines.

With the contracts signed, Factory 36 commenced construction of two new production buildings, which were due for completion at the end of 1978. Factory 135 also started preparing for the assembly of the GEM engine. Westland meanwhile sent 20 of a total of 70 technicians earmarked for the project to Cairo to commence operation. Before further work could take place, however, the start of the Camp David process led to a suspension of the AOI, and thus of its various activities. In the case of the *Lynx* programme, this led to mutual legal action by Westland and the Egyptian government, which has yet to be resolved. But with restoration of Arab-Egyptian ties and the prospect of a revival of co-operation through the AOI at the end of the 1980s, talks with Westland were resumed. The company is now offering local assembly of the EH-101 helicopter in the version produced by Italian Augusta, as acceptable compensation for loss of the *Lynx* contract. This would be run as a joint venture worth some $1bn, with Rolls-Royce engines also being produced by ABECo.[104]

• *Gazelle*: following the collapse of the *Lynx* programme, and faced with its own continuing need for a light utility helicopter, Egypt initiated a new programme for the local assembly of the French Aerospatiale *Gazelle*. The Egyptian Air Force already had 54 SA 341 *Gazelles* in service, which had recently been retrofitted to SA 342L standard by Aerospatiale, and in December 1981 it placed a follow-on contract for 36 more. Of these, 30 were to be assembled by ABHCo, beginning in June 1983. The first locally-assembled helicopter was delivered in December 1983, though Aerospatiale was really responsible for the hand-over, since ABHCo was in fact acting as its sub-contractor. This meant that it was French technicians who undertook quality control and testing.

Initially, the production rate was two *Gazelles* monthly, rising to three by February 1985, though maximum factory capacity was four.[105] Assembly was from French-supplied kits in which all major components were already assembled. ABHCo hoped nonetheless to assemble the transmission and to learn composite-materials technology for the components. Meanwhile, ABECo was undertaking assembly of the Turbomeca *Astazou* XIVH engine for the helicopter, and by programme end it had produced 48 (including 18 spares). Furthermore, by February 1983 ABECo was already doing engine repair work at a new repair/test/overhaul facility.[106] By the beginning of 1988 production of *Gazelles* was complete.

• *Super Puma* and others: in November 1983 the AOI and Aerospatiale signed a protocol for local manufacture of components for the French medium transport *Super Puma* helicopter.[107] This was essentially a sub-contract in which Egypt supplied Aerospatiale production lines. The AOI hoped to move eventually to assembly of the complete helicopter, but lack of funds has prevented this so far.

Towards the end of the 1980s, two other VTOL aircraft projects came into consideration for local production in Egypt. These were the AH-1W *Super Cobra* helicopter produced by Bell Helicopter Textron, and the V-22 *Osprey* tilt-rotor aircraft under development by Bell-Boeing.[108] In both cases the American companies looked to

Egypt as a useful base for regional sales; an added advantage was that Egypt could use its FMS Foreign Military Sales credits to back these projects, yet remain in conformity with FMS programme conditions, as payments would be to American (subsidiary) companies.[109] As the V-22 *Osprey* programme was halted by Congress in early 1989, it is no longer a practicable proposition. However, the *Super Cobra* – Bell AH-1W plus Viper – is the most likely contender to win an EAF requirement for a dedicated combat helicopter (other candidates are Agusta A-129 *Mangusta*, Bell AH-1S *Cobra*, and Hughes AH-64 *Apache*).[110]

A final potential helicopter project is the EH-101. This is being offered by British EH Industries (part of Westland) and Italian Agusta as a means of resolving the decade-old dispute between Egypt and Westland over the cancellation of the *Lynx* contract following the suspension of the AOI in 1979. The proposal will include local assembly of the EH-101.[111]

RPVs

Despite the adequacy of its aerospace and electronics industries, Egypt has not yet presented its own RPV, although Air Defence Commander Adil Khalil announced in October 1987 that five versions were under development.[112] Nor is production or assembly under licence known to be taking place, although several projects are under consideration. One of these is the offer made in late 1987 by Aeritalia for co-production of RPVs, as part of a wider package also including production of the G.222 tactical fighter and avionics equipment.[113] Since then, the AOI has considered producing the Austrian PH-23 RPV or the British *Optica*, though neither has been confirmed.[114] Indeed, Egypt moved instead, in March 1988, to order the American-built *Scarab* RPV.[115] Whether this indicates a change in policy away from local designs remains to be seen.

Ground-Based Weapons

As in most developing countries, the limitations placed on military infrastructure and manpower by prevailing technical and educational conditions have tended to compel the Egyptian armed forces towards an emphasis on ground forces. Similarly, despite constant attempts to raise technical standards generally and develop specialised services such as the Air Force, a large part of equipment remains relatively unsophisticated technologically.

Additional factors affecting the procurement or local production of ground weapons have been financial constraints and combat experience. The former factor has limited the types of weapon system and associated technology that Egypt has sought to acquire or replicate. The latter factor – specifically the experience of the October 1973 war – demonstrated the extraordinarily high rates of consumption of ordnance and equipment in modern warfare, and so pointed to the need for large stockpiles of basic items. It also highlighted the need to reduce dependence on foreign supplies during wartime, and the possibility of cutting costs through local production.

Against this background, and with the added incentive in the 1980s of the possibility of exporting local military products to other developing countries facing similar conditions, it was natural that the production of items for the ground forces should be the longest-standing activity of the Egyptian defence industry. Indeed, this line has an unbroken record since the late 1940s, in contrast to the aerospace industry, which was effectively suspended for most of the 1970s despite an early and continuing interest and major investment. Production activity in the area of ground weapons has been divided among several main categories.

Armoured Vehicles

The most immediate reason for Egyptian interest in armoured vehicles, especially tanks, was the desire to match the development of Israeli armament in the early 1950s. This was reflected in the conclusion of the Egyptian-Czechoslovak arms deal of 1955, by which Egypt acquired the T-34 tanks that were to see action in the Suez war of 1956. The Egyptian Army was impressed by Israeli use of all-tank formations and mobile tactics during the latter campaign, and sought new T-54/55 tanks and anti-tank weapons as a result. The importance of mobile warfare and of armoured vehicles was emphasised once more by the stunning Israeli success in June 1967, with the fall of the Sinai Peninsula (as well as the Gaza Strip, the West Bank, and the Golan Heights) in a matter of days to the Israeli Army's blitzkrieg and 'conveyor belt' tactics.[116] Constantly out-gunned and out-manoeuvred by Israeli tanks, the Egyptian Army moved after the 1973 war to upgrade its own and to further integrate its combined armour-artillery-mechanised infantry teams.

Despite the importance of armoured vehicles, little was done to produce them in Egypt until the 1980s, with the sole exception of the *Walid* APC, which has been produced locally since the late 1960s. But political and financial considerations have changed the picture. The re-orientation of Egypt's foreign policy under Sadat towards the West and the USA, especially from 1979 onwards, resulted in the introduction of new items to the Egyptian inventory that had consisted overwhelmingly of Soviet-designed equipment until then. The willingness of Western countries to allow licensed production, coupled with the Egyptian need to integrate Western and Soviet weapon systems into service, formed a major incentive for the development of local capabilities for modification and manufacture. At the same time, the increasing cost of arms imports compelled the Egyptian armed forces to seek ways of prolonging the operational life of out-dated equipment. Previously it was cheaper to buy new Soviet hardware than to attempt to produce it locally, but the collapse of Soviet-Egyptian military relations and the high price tags on Western products altered calculations.

Egyptian efforts have therefore been focused since 1983 on three main projects. These are the upgrade of T-54/55 and T-62 tanks, production of a new tank such as the American M-1 *Abrams*, and the development of the *Fahd* APC. In the case of the M-1 specifically (or of any other tank design chosen for local production), Egypt is seeking to acquire advanced technology and manufacturing experience that will serve as the basis for the establishment of a viable, armoured-vehicle industry, leading eventually to the growth of local design and production capabilities. Similar aims are evident in such activities as uparmouring American-made M-113 APCs, and in the (unimplemented) plan to produce the British Alvis *Scorpion* light tank locally.

At present, however, the role of imported technology is overwhelming. This applies not only to the proposed production under licence of the M-1, but also to the T-54/55 upgrade programme. In the latter none of the sub-systems being substituted are of local design, though several are licence-produced. More significantly, the major part of the redesigning involved has been undertaken by foreign companies, as well a part of actual implementation. Only the *Fahd* APC can be considered as the result of indigenous design and development, though its principal components (engine, transmission, and chassis) are of imported design or manufacture.

• *Tanks: the T-54/55 upgrade*: at the beginning of the 1980s, the Egyptian Army had an estimated 850 T-54/55s in service, representing half its total tank inventory.[117] Originally designed in the 1950s, a large number of this medium tank were delivered to Egypt in the decade up to 1973. By 1980, therefore, the T-54/55 was both seriously outdated compared with other AFVs in service in the region and also approaching the end of its operational life. Earlier in the 1970s, some upgrading had been undertaken by adding an AEG Telefunken white/IR searchlight and, on some tanks, an *Iskara* laser range-finder. Faced with severe financial constraints that prevented complete replacement with new imported vehicles, the Egyptian

Army adopted an extensive upgrading programme aimed specifically at bringing its T-54/55s up to the standard of the American-made M-60A3 tank also in its service.[118]

In effect, the T-54/55 upgrade programme has run along three parallel courses. In 1983, the Egyptian Defence Ministry announced that it had agreed with Romania to produce under licence the latter's version of the T-54/55 – the M-77. According to the deal, Egyptian factories would also produce spare parts for some weapons and electronic systems. The main changes involved were the addition of a sixth road-wheel to the running gear, and the addition of side-skirts for protection against hand-held, infantry AT weapons.[119] However, the upgrade was deemed insufficient for local needs and Western sources suggest that the programme was probably abandoned.[120] Nonetheless, Iraq was known to have taken possession of 250–260 M-77s, which reportedly were assembled in Egypt.

As the Egyptians were dissatisfied with the old 100 mm gun of the T-54/55, they planned to replace it with the British-designed 105 mm L-7/M-68, as a separate modification. A contract worth $10m was accordingly signed with Britain's Royal Ordnance Factories (ROF) in June 1985. Under its terms, ROF was to provide retrofit kits of the L7A3, the version it had already developed for installation on T-54/55s, involving a 180° turn in the loading position and a revised recoil system, as well as other modifications in the turret, fuel, and ammunition storage tanks and in the method of gun removal.[121] The gun system was to be manufactured in Britain and fitted in Egypt, with deliveries commencing in 1985.

These alterations were inadequate to improve the capabilities of the T-54/55 radically, however. The Egyptian Army therefore initiated a second upgrade programme, with the additional aim of standardising parts among tanks in its service. This involved replacement or introduction of new night sights, aiming devices, signals equipment, and infra-red systems, as well as the installation of the L7 gun. Previously, many of the T-54/55s had been refitted locally before 1973 with the German AEG infra-red/white light searchlight and the Yugoslav *Iskra* laser range-finder, which were now replaced.

The engine remained the same, however. So did the suspension and fire-control system, owing to financial restraints, although they were supposed to be replaced or added. Work on the first 160 T-55 tanks – Phase I – was to be undertaken by Factory 100 and the workshops of the 2nd Army, which had 33 prototypes under production by Spring 1987. Parts and components were delivered by ROF, though the second batch of 100 T-55s were supplied with parts by Abu Za'bal. This would eventually leave some 150 T-55s with only minor modifications (mainly the gun). Full manufacture commenced in July at the rate of 12 monthly. According to the Minister of State for

Defence Production, Jamal as-Sayyid, production costs here were 30 per cent lower, saving E£31.2m.[122]

Evidently, despite claims by then-Defence Minister, Abdul-Halim Abu Ghazaleh, that the new version T-54/55 was '100 per cent Egyptian-made', the element of imported components and technology remained overwhelmingly predominant.[123] Local contribution was limited to the value added in the form of labour used in refitting or in cut-steel components (at Abu Za'bal) and the licence-producing of some of the new sights (at Arab Optics International).

Well before this last programme was underway, Egypt had already launched its third parallel course of upgrading the T-54/55, as a result of the inadequacy of the Romanian M-77 version. Late in 1984 the American company Teledyne-Continental Motors was contracted to develop a new version of the T-54 – also designated *Ramses*-II. The 450 T-54s still in service were seriously aged, requiring a complete retrofit. The design offered by TCM involved a thorough remanufacture, with replacement of the engine, transmission, suspension, gun, fire-control system, and all other items short of the actual hull and turret.[124] Fire suppression and NBC systems were also added to a prototype T-54 which was shipped to the USA in March 1985. The improvements led to a better power-to-weight ratio and to superior gun accuracy and lethality, with testing being conducted in Egypt after the return of the first vehicle in late 1986. According to the agreement, TCM would supply 100 retrofit kits for assembly in Egypt.

Despite the work done by TCM, the Egyptian Army opened the door for additional bids, and received a new offer from a European consortium headed by ROF in March 1987. Both TCM and ROF exhibited their designs at the Cairo International Arms Exhibition in November that year, as did Austria's Noricum. Following further development and tests, the version offered by ROF was finally approved in April 1988. While the latter provided overall design and the main armament, the German MTU and the British Horstman Defence Systems companies were to provide the key sub-systems, such as a Mercedes-Benz diesel engine coupled to a ZF fully-automatic transmission in order to improve the power-to-weight ratio enough to compensate for the increase in total weight from 36 to 42 tons.[125]

This initial decision for a test version notwithstanding, the Egyptian Army had not taken a final production decision by summer 1989, although it was interested in buying 150 retrofit kits from ROF. In the meantime, TCM remained a possible candidate, with a potential contract to provide 425 kits over six years, followed by Egyptian manufacture of kits for local use and re-export.[126] With Egyptian moves towards setting up a tank assembly plant for a modern MBT well underway, the constant delays in the T-54/55 upgrade programme suggest that its more ambitious aspects may be dropped altogether. Instead, after years of trials and partial modifications (since 1983) the Army may be satisfied with the replacements it has already made of the gun, main sights and fire-control systems, until new tank types enter service later in the 1990s.

• *Tanks: the T-62 life extension*: among the Soviet tanks in Egyptian service are some 600 T-62s. First entering production in 1961/1962, it is sufficiently improved over the T-54/55 as to retain greater combat viability. The Egyptian Army is moreover reasonably satisfied with the performance of its 115-mm smoothbore gun, for which shells are produced locally. However, age and use have necessitated replacement of the main armament, with ROF manufacturing a number of 115-mm guns for Egypt since 1985. Austria's Noricum mounted an L7 105-mm gun on one T-62 test-bed, but this option was rejected by the Egyptians, though it was later exhibited in Baghdad in May 1989. (A proposal by GIAT to mount a 120-mm smoothbore gun remained on paper only.) It is assumed that certain sub-systems (sights, infra-red devices, and fire control) have been replaced locally, but no further upgrade has been reported. Previously, some T-62s in Egyptian service had been provided in the 1970s with four turret-mounted 80-mm launchers (D-3000 or D-6000) firing smoke rockets, and with AEG Telefunken white/IR searchlights and *Iskara* laser range-finders.

• *Tanks: the* Scorpion *interlude*: the Romanian T-54/55 (M-77) was not the only tank for which Egypt obtained a production licence in the 1980s. As early as 1982 an Egyptian defence industry company had expressed an interest in producing a light tank. The company was Arab Electronics International AEI (precursor of Arab Optics International), and the tank was the *Scorpion* designed by Britain's Alvis Company. Negotiations were conducted between and United Scientific Holdings (Alvis's mother company) in April 1982, but a decision was taken only in 1985. By then both AEI and USH had come together to form the Arab Optics International joint venture, which was granted a licence to assemble the *Scorpion* locally. This was not taken up, however.

• *Tanks: Project 200 and the M-1* Abrams: Egyptian officials first indicated their intention to produce a new tank (Project 200) – rather than upgrade an existing one in their inventory – in 1984, when the Defence Ministry announced plans to establish a manufacturing plant for the purpose. Later in the year General Dynamic Land

Systems was contracted to undertake the construction, following apporval by the government of the plan. The total cost of the project was set at $720m (£600mn at current prices), of which $150m was the value of the construction contract. Initial development was to receive $350m, of which the armed forces allocated $216m (£177.05m) from their 1984/1985 budget, with an additional $90m (£75m) coming from the Ministry of Defence Production.[127]

According to the original plan, the new tank plant – assigned the title of Factory 200 and operating under the NODP – was to come into operation in 1987. Initially it would provide overhaul and maintenance facilities for M-60A1 tanks and M-881A1 armoured recovery vehicles (engines and transmissions, especially). At a later stage, a new tank of imported design would be produced there under licence, after a phase in which complete vehicles would be purchased from the original supplier. (Production capacity was set at 120 tanks yearly.) In the meantime, the plant would also undertake renovation work on tanks in the current Egyptian inventory.[128] Explaining one of the reasons for producing a tank locally, Abu Ghazalah stated that it would be 30 per cent cheaper than an imported version;[129] as the type had not yet been chosen, this was an obvious reference to expected savings in labour costs. At this point, the only specification was for a 46-ton tank: this automatically excluded the American M-1, the German *Leopard*-2 and the British *Challenger* and Vickers-7.[130] The last tank came under consideration in 1986, though, as did the Brazilian EET-1 *Osorio*, the Italian OT-40, and the French AMX-40, with view for an initial order of 96–100 vehicles and a final total of 400.[131]

While the Egyptian Army studied its options, construction of Factory 200 proceeded, and its Tank Depot No. 1 was completed in January 1988. This could now provide support for the computerised fire control system on the M60A3 tank, and a test depot was installed by Kollsman (US) at a cost of $5m (but separate from Factory 200).[132] Sited near Factory 100 (producing parts for M-60 overhaul and maintenance), it covered an area of 2.7 square kilometres, of which the main plant occupied 80,000 square metres.[133] At 600 metres long, the Egyptians proudly claimed that it was the largest building in Asia, Africa, or Europe. It lacked a turret-making facility, though, which meant that in any new project turrets would have to be manufactured by welding rolled steel rather than cast, with attendant structural weaknesses.[134]

In parallel, the Egyptians set up steel-producing lines to feed the new tank plant. The first step came in mid-1986, when a joint project with the Egyptian Iron and Steel Company was planned, using a British loan.[135] As Minister of Defence Production Jamal as-Sayyed explained, this activity would be undertaken by the civilian Ministry of Industry, as military need was one-tenth as large and so could be undertaken more economically in the civilian sector.[136] However, this effort faltered, as Egypt was not in a financial position to use its line of credit. Similar problems continued to delay production plants for carbon steel, alloys, titanium oxide, and other materials; by 1990 both Egyptian-Arab and Egyptian-Korean joint ventures were still being hindered by the inability to raise local capital to meet the required 51 per cent quota for Egyptian input.

Despite going ahead with the construction of Factory 200, it took Egypt two years to make a final, irrevocable decision on which tank to produce there. Having specified originally that it wished to produce a tank in the 46-ton range, the Egyptian Army moved firmly by spring 1987 to prefer the American-designed M-1 *Abrams*, although the latter weighed some 55 tons.

The change in goals was propelled by the powerful Defence Minister, Abdul-Halim Abu Ghazalah, with whom the M-1 and other defence projects have been identified. His personal role also appears to have been the reason for much confusion surrounding the project, as his statements in support of producing the M-1 often conflicted sharply with those of American and other officials. As if to underscore this fact, following Abu Ghazalah's dismissal in April 1989, Egypt considered dropping the M-1 altogether, and even considered the Soviet-designed T-72 as a possible replacement. In any case, production of the M-1 finally went ahead only in mid-1991, nearly two years after its original timetable.

In essence, the debate over whether or not to produce the M-1 demonstrates most clearly the nature of Egyptian ambitions for the defence industry and the basic underlying financial, technical, and political calculations.

An early indication of the extent of the change in Egyptian plans, and in Egyptian ambitions, came in June 1987 when Abu Ghazalah revealed that Project 200 aimed to produce between 1,000 and 1,500 M-1s (compared with an original target total of 400 vehicles).[137] Funds would be provided partly by diverting $168m from American MAP assistance for 1988; and, in fact, a major part of Egyptian credits would have to be diverted for several years, affecting other weapon programmes. This meant allowing Egypt to receive part of its annual aid in cash and spend it outside the USA in contravention to the standing conditions of MAP, for which it obtained President Reagan's initial approval. There was an immediate precedent: Israel had twice been permitted to divert part of its MAP allocations (over several consecutive years in each case), for the development of its *Merkava* tank (1973–78) and *Lavi* combat aircraft (1983–88).

Immediate protests were voiced by Israeli supporters in the US Congress and by some other parties. Expressing the general argument against granting Egypt a production licence, a *New York Times* editorial on 6 July pointed out that the M-1 contained the latest in American technology, much of which was moreover classified, and that it was therefore highly unsuited for the Egyptian armed forces as well as inappropriate for the Egyptian defence industry. For the price of each M-1, conversely, Egypt could buy 3 M-60s, which were better suited to Egyptian capabilities and more sensible in view of Egypt's critical economic situation.

That serious financial, political, and technical issues were indeed at stake was evident in statements by Abu Ghazalah and other Egyptian officials about the M-1 project. The core questions were the following: total cost and means of finance; the work share between Egyptian and American plants; the transfer of technology; and the split between local assembly and production.

On the first count, Egyptian officials indicated that they expected to use a substantial portion of their annual MAP allocation of $1.3bn for several years, or at worst to obtain funding under the United States' FMS programme.[138] Total cost was set at $2bn, though to produce Abu Ghazalah's ambitious aim of 1,000–1,500 tanks would in fact raise the final bill closer to $5bn.[139] Defending the M-1 option, the Egyptian Defence Minister and other industry officials insisted that it could not only be produced at a commercially viable unit price (compared with American prices), but at even lower cost than in the USA owing mainly to lower labour costs.[140] However, American sources insisted equally firmly that it would cost Egypt $4m to produce each M-1 when all costs had been factored in (initial investment – plant, tooling, training, etc – development, and production), compared with $3m per unit when imported under FMS or $2m per unit supplied to the US Army.[141]

Despite these arguments, negotiations continued, and in April 1988 the Pentagon formally notified Congress of its approval of licensed production of the M-1 in Egypt. Congress ratified the programme, and a Memorandum of Understanding (MoU) was signed by US Secretary of Defense Frank Carlucci while visiting Cairo in November. According to the agreement, 15 completed M-1s would be delivered to Egypt in 1991 – although Abu Ghazalah had earlier asserted that the first delivery would take place in 1989 –[142] following which assembly would commence at Factory 200 of the first of another 540 tanks. Production would go through six stages, broadly divided into: assembly using kits provided by the USA; local production of parts, such as sprockets, road wheels and tracks; then additional local co-production – including re-export of components to the USA – reaching a maximum ceiling by the end of serial production in 1996.[143]

These terms contained additional points of contention, however, leading to major confusion between Egyptian and American pronouncements. A minor point was that the delivery date for the first locally-assembled M-1 was 1992, at least two years later than predicted by Abu Ghazalah to the public. A far more serious issue, though, was disagreement over the American-Egyptian work share. For domestic political reasons, Abu Ghazalah was anxious to stress the economic benefits to Egypt of producing the M-1 despite its high costs. Thus it was claimed that Egypt's final share in M-1 production was to reach 40 per cent, a figure promptly denied by American officials, who set the correct share at 19 per cent.[144] Of the $2bn programme cost, $1.6bn was to go to American companies in direct and indirect income, according to the conditions set by the Pentagon and Congress in Spring 1988.[145]

A similar point of serious contention, with political and strategic implications for both sides, was the issue of technology transfer. In summer 1987, Egyptian sources confidently confirmed that the USA would probably waive its 'national disclosure policy' if Egypt committed itself to safeguarding secrets such as metallurgical processes for armour.[146] However, in approving licensed production, Congress specifically prohibited the transfer of classified components. The MoU finally signed in late 1988 indeed excluded such technological areas as the engine, transmission, armour, and the fire-control system from the 19 per cent work share allocated to Egypt.[147] Egypt was moreover not allowed to make third-party sales without American approval; the prospect of exporting the M-1 had formed an argument used by Abu Ghazalah in favour of adopting it, giving rise to his production goal of '1,000 to 1,500' tanks. As if to underscore this point, the USA moved in September 1989 to sell 315 M-1s to Saudi Arabia, for a total contract value of $1.5bn.[148]

The cumulative effect of this conflict in interests was to provoke a constant tug-of-war between the two sides. An instance is the struggle over the role of the General Motors subsidiary in Egypt. Seeking additional, indirect, commercial benefit, the American negotiating team pressed for a greater work share for GM-Egypt, which the Egyptians resisted, partly on the grounds that the latter company was only willing to undertake assembly rather than local production.[149] They also hinted that they were considering both suspending the annual joint manoeuvres with American forces known as '*Bright Star*' and replacing M-1 production with the Soviet T-72.[150] Significantly, the latter leaks came after the dismissal of Abu Ghazalah, who was generally considered to be both an American ally and more amenable to waive core points, such as technology transfer.

The leaks also preceded the visit to Cairo in 1989 by US Deputy Secretary of Defense Richard Armitage, who arrived to discuss

American assistance to Egypt and the payment schedule for the latter's massive military debt. Among other issues on the agenda with President Mubarak was the M-1; Armitage agreed to the diversion of $450m of MAP funds into cash for Egyptian use (on condition that total military expenditure was reduced).[151] With this, differences over the M-1 project appeared to be resolved. In March 1990, the Defence Production Ministry assigned contracts to construct production lanes at Factory 200, but it took until the beginning of 1991 before bids could be invited to supply the machinery for the plant.[152] Production of the first locally assembled M-1s was supposed to start in the second half of the year. Egypt was the only country outside the USA to produce the tanks, and, indeed, is set to become the only country to produce it anywhere following the American decision to cease M-1 production.

- *APCs*: Walid: the *Walid* was Egypt's first locally-produced APC, using a design very similar to the Soviet BTR-40. The Egyptian Nasr Automotive Plant at Helwan commenced series production in the late 1960s, providing a total of some 1,000 four-wheeled vehicles to the Egyptian Army by the 1980s. An additional number were exported to Algeria, Sudan, North Yemen, Guinea, and Burundi. Based on a German-designed chassis built under licence from Magirus-Dietz, the *Walid's* armoured body was also constructed at Nasr, though assembly was undertaken by Qader. Production of an updated version – *Walid* 2 – was started in 1981, based on Mercedes-Benz automotive parts. It has probably stopped completely since then, with effort devoted instead to the new *Fahd* APC.

- *APCs: M-113*: before moving to develop an indigenous vehicle, another production plan mooted by defence industry officials in an early phase was that of the American M-113 tracked APC, which Minister of Defence Production as-Sayyed said was under consideration in 1980.[153] Little came of this particular proposition, though the Qader factory later discussed converting the M-113 into a mortar-carrier with Western companies.[154] Eventually production plans were dropped completely, though a modification has been proposed based on locally-designed add-on armour, consisting either of steel plates or of a lighter alloy to retain buoyancy.[155] The version developed by Qader Factory consists of 5–20-mm thick steel plates sandwiched between two layers of rubber, and provides protection against 23-mm projectiles at 100m. Buoyancy is retained by placing foam rubber between the add-on armour and the vehicle; the weight increase is 650 kg.[156]

- *APCs*: Fahd: based on the experience gained in producing *Walid*, the Egyptian defence industry moved in the early 1980s to develop a new APC. This was the four-wheeled *Fahd*, based on a Daimler-Benz LAP1117/32 truck chassis with a locally-produced armoured body. The first prototypes (TH390) were designed and built by Thyssen Henschel and delivered to the Egyptian Army in Spring 1984. Qader Factory undertook the work, using a Daimler-Benz LAP 1117/32 chassis.[157] In fact, *Fahd* represents a hybrid of local and imported design. The engine is a water-cooled, Daimler-Benz OM 352A turbo-charged diesel coupled to a Daimler-Benz manual transmission, while Qader collaborated with Germany's Thyssen-Henschel to develop the armour. Qader also provided the overall design, while the turret (for the version armed with a roof-mounted machine gun) is designed and produced by Sakr Factory. The end-result is an 11-ton vehicle with 800-km range, maximum speed of 90 km/h on the road and 50 km/h cross-country (but no amphibious and little fording ability).

Since final testing in summer 1985, several hundred *Fahd* APCs have been produced. Of these an estimated 200 were in service with the Egyptian Army by 1990, while additional numbers have been exported to Oman, Zaire, and Kuwait, making it one of the more successful local defence products. For the Kuwaiti consignment – part of a $50m deal signed in spring 1988 for 100 vehicles – Qader purchased Western-made cooling systems, while the vehicles sent previously to Oman and Zaire were supplied with imported Creusot-Loire and SAMM turrets.[158] In order to increase *Fahd*'s versatility and enhance its attractiveness to foreign customers, Qader is planning a family of vehicles, including command, reconnaissance, ATGM/MRL launcher, recovery, and ambulance versions.[159]

Artillery

It was only at the beginning of the 1980s that serious effort was put into producing artillery weapons locally, as the Egyptian defence industry relaunched itself. Unlike such categories as small arms or even the *Walid* APC, the availability of cheap Soviet weapons and the technological complications involved in certain production aspects (such as boring and lining gun barrels) had discouraged local efforts previously. Interest in producing artillery systems locally was also spurred by the first Gulf war. Iraqi need for artillery weapons for use against Iran was met by massive Egyptian exports; these were taken from Egyptian Army stocks, creating a shortfall. With income from exports, the defence industry had an opportunity to develop its manufacturing capabilities and offer improved systems both to the Egyptian Army and to foreign customers. An added incentive generally was that the Gulf conflict, and later Israeli use of specialised artillery munitions during the invasion of Lebanon in 1982, reiterated Egypt's own belief in the importance of massed, ground-based firepower.

The main effort of the Egyptian defence industry in this area since 1980 has gone into producing or upgrading existing gun and rocket artillery systems in service with the Army. The emphasis has therefore been on Soviet-designed weapons, though innovations have been sought in developing advanced warhead munitions and hybrid self-propelled systems. As a result, there has been less dependence on importing technology in this field, with greater resort to reverse-engineering and local R&D. Nonetheless, there has been regular co-operation with other countries, such as China and Britain, in the production of gun barrels or the development of self-propelled systems (especially the vehicle mounts).

• *Artillery: Gun Artillery*: despite having initiated local manufacture of the Soviet-designed 130-mm gun (the M59–1M with a range of 27.15km), the absence of potential customers has since prompted the Egyptian defence industry to focus solely on the 122-mm howitzer, also of Soviet design. After conducting technical studies in 1980, manufacture was supposed to commence the following year in co-operation with British firms.[160] In the event, however, the D-30 model of the 122-mm howitzer was produced locally through reverse-engineering, though with technical help from China for such specialised processes as barrel manufacture. Thus it was only at the end of 1984 that an early production version was exhibited by Abu Za'bal Engineering Industries Co. (which also produced the M59–1M). According to Egyptian officials, local capacity was for 300 122-mm guns annually, with the USA also expressing interest in buying the D-30.[161] In addition, the Helwan Machine Tools Co. produces a copy of the Soviet M1943 120-mm mortar.

• *Artillery: Self-Propelled Artillery*: in line with the general movement of modern armies towards mobile artillery capable of accompanying armoured units in combat, Egypt has sought to increase the self-propelled (SP) component of its artillery arm. For the most part this has meant importing American-made 155-mm M-109s. The Egyptian Army has also sought, though, to develop a new system based on its 122-mm howitzer, in order partly to utilise local production capability and partly to retain an effective weapon.

The first practical step towards the latter end came in 1982, when the Egyptian Ministry of Defence Production signed a protocol with Britain's Royal Ordnance Factories for the development of a 122-mm self-propelled howitzer. Soon after a competition for bids was opened, with the American company BMY submitting a second proposal. In both cases, the actual Egyptian input was very low, consisting of little more than the gun itself. Overall design for the hybrid was to be by the foreign companies; it had to accommodate for the complete 122-mm howitzer (minus road wheels and trails), according to the Egyptian Army requirement. The original requirement was for 700–800 SP guns, though financial restraints eventually reduced this considerably.

By early 1985, both BMY and ROF had submitted prototypes, as had Abu Za'bal Engineering Industries Co. The ROF design, which was eventually preferred, was based on a modified chassis of the British Combat Engineer Tractor, with a view to developing a family of armoured vehicles. Unlike the BMY proposal for using an aluminium alloy, ROF used steel in order to facilitate production of its SP122 in Egypt.[162] Thus the Egyptian defence industry would increase its input from simply providing the gun, by participating in construction of the hull and in assembly. Initial demand was for 200–300 systems, but by 1990 only a handful were in service, casting doubt on the exact status of the programme.

• *Artillery*: *Anti-Aircraft Artillery*: Egypt possesses large numbers of Soviet-designed and -supplied anti-aircraft guns of several calibres. In order to prolong the usefulness of some of these – particularly the 23-mm gun – the Egyptian defence industry has sought to modernise them by providing them with up-to-date fire control. This applies principally to the *Sinai*-23 and the *Amun*/Contraves *Skyguard* system (both described below), and to the towed, twin-barrelled 23-mm weapon. The latter is manufactured locally as the ZU-23M. The project is undertaken by Switzerland's Contraves on behalf of the Egyptian Ministry of Defence Production, according to a deal signed in early 1985. The system consists of up to four guns slaved to a single *Gun King* computer-controlled sight. The contract calls for co-production and common marketing.[163]

• *Artillery*: *Rocket Artillery*: rocket artillery has consistently occupied a prominent position in Soviet-supplied Third World armies, the Egyptian one being no exception. Following the near-total suspension of military supplies from the USSR, Egypt moved to manufacture and modify its rocket artillery weapons. Of these, the 122-mm system is the most important, as it is the most widely-used, and has thus had the most modifications.

Starting in 1980, accordingly, development started on a new 122-mm rocket. This was undertaken by Saqr, in conjunction with French and German sub-contractors.[164] Testing commenced in May 1981, followed by production in 1983, in which the same European aerospace partners also took part. An additional incentive for production in this period was the Iraqi demand for 122-mm systems – both

BM-21 launchers and ammunition – which Egypt supplied both from inventory and new production. Among the main modifications to the rocket were an electronic fuse developed by Saqr, though a major improvement was the increase in range to 30 km, giving it the name of *Saqr*-30. Another version of the 122-mm rocket – the *Saqr*-18 – was provided with new warheads of local design containing special sub-munition dispensers: 28 anti-personnel (lethal radius 15m) or anti-tank (80-mm penetration) bomblets, five anti-tank mines, or 17.5 kg of high explosive. The range of *Saqr*-18 was 18km, however. Generally Soviet-supplied BM-21 launchers or locally-assembled banks of tubes were used with the new rockets, though by late 1987 the Helwan Machine Tools Co. (Factory 99) was producing a multiple launcher based on a 6×6 chassis, as well as light one- and three-round, tripod-mounted launchers for use by commandos (the PRL-111 and PRL-113, with a range of 3–10.8 km).[165] The MRLS comes in two versions: 30 tubes (two units of three, five-tube banks each) on a Zil-131 or Japanese Isuzu truck, or a 40-tube version on the Zil.

Since then, Saqr has developed a number of additional rocket artillery systems, not all of which are based on older Soviet designs. A hybrid is the old 132-mm system, which is now produced as a 16-tube multiple launcher. A new system is the VAP 80-mm multipurpose rocket, of local design. Mounted on a jeep in a 12-tube configuration (with two stabilisers), this has an 8-km range and can be used (among other things) as a target for anti-aircraft training.[166] A second 80-mm weapon is the 12-round D-3000 mounted on the *Waleed* APC, which can lay a smokescreen 1,000m wide and lasting 15 minutes at up to 2.5-km range. The system can also be mounted in twin 2-tube fittings on the T-62 turret, as can the heavier D-6000 version with its 7-km range.

- *Artillery: Tactical Ballistic Missiles*:

▶ the *Zafir*, *Qahir*, and *Ra'id*: as part of the early, ambitious phase in Egypt's military industrialisation under Nasir, a special programme was established in 1960 to develop heavy, surface-to-surface bombardment missiles, with a single-stage and a two-stage missile being first tested and then paraded publicly in July 1962. Development of these weapons was the responsibility of the Heliopolis Missile Plant, which designed three types: *al-Zafir*, *al-Qahir*, and *al-Ra'id*. Some 100 missiles of all types were produced. *Zafir* and *Qahir* carried warheads of 450–500kg and 670kg to distances of 375–390km and 600–620km, respectively. The third – *al-Ra'id* – was assumed to be for non-military purposes in fact and capable of carrying either a 180-kg warhead to 900–950km or a 1,000-kg scientific probe to 700km. These performance data remained largely hypothetical, however: accuracy was poor owing to an inefficient inertial guidance system, and operational warheads had probably not been successfully developed by the end of the programme.

These weapons were nonetheless the pride of the Egyptian defence industry, although they were in fact designed with the assistance of a group of 80 German scientists (supervising) and some 600 other Western technicians.[167] During the five years of activity before the programme was suspended, these advisers assisted in the formation of a team of 150 Egyptian engineers and technicians, and in the setting up of a number of associated R&D bureaux.[168] It is possible that some development work on the *Zafir* and/or the *Qahir* missile has been revived at Saqr – since a few are rumoured to remain in service – especially to provide them with more advanced guidance systems, though the reports may in fact be referring to duplication of the Soviet *Scud*-B.

► *Scud*-B: originally supplied by the USSR between 1967 and 1973, the Egyptian Army is reported to have 9 *Scud*-B systems (range 280 km) still in service, with an indeterminate number of missiles.[169] In order to reinforce this arm of service, the Egyptian defence industry may have produced a domestic copy, with North Korean aid, though this is unconfirmed.[170] Other reports state that, in fact, Egypt has helped Iraq to copy the *Scud*-B, or even to modify it to extend its range as the Iraqi *al-Husein*.

► *Saqr*-80: the most impressive confirmed Egyptian SSM development, though, is the *Saqr*-80. First revealed in late 1987, this rocket can carry a 200-kg HE warhead to a distance of 80 km. It can be loaded with a submunitions dispenser which releases either 1,000 AP/AT bomblets or 65 anti-tank mines.[171] Some 20 systems are now in service in Egypt, and a number have been sold to Iraq, which is rumoured to have participated in developing *Saqr*-80.[172]

► *Condor*-2/*Badr*-2000: although little hard information is available, and despite repeated denials by officials in the countries concerned, it appears that during the 1980s Egypt and Argentina were involved in developing jointly a new medium-range ballistic missile known alternatively as *Condor*-2 or *Badr*-2000. Iraq is also believed to have provided finance for the project – though it too has denied this – and to have taken over development at its own facilities.[173] Severe American pressure has led since then to the cancellation of Argentinian and Egyptian efforts: the USA confirmed Egypt's withdrawal in September 1989, though President Mubarak countered by stating that local contribution to the programme had stopped a year earlier.[174] Argentina confirmed complete cessation of its role in mid-1991.

- the *Badr*-2000 is thought to be a solid-fuel weapon capable of carrying a 450-kg payload to a range of 800–1,000km, potentially placing the whole of Israel (and Libya) within Egyptian range.[175] In June 1988, American federal agents arrested five people, including an Egyptian military procurement officer, in Washington and an Egyptian-born scientist working at an American defence firm on charges of attempting to smuggle high technology materials to Egypt. Among the items were know-how on working with carbon-carbon, stealth technology, rocket-exhaust nozzles, and microwave telemetry antennae. Together they formed 'a complete package to build or upgrade a tactical missile system'.[176] This incident is thought to have contributed to the dismissal of Abu Ghazalah, who had authorised the smuggling operation, but it is not known if it was meant to assist the *Condor*/*Badr* programme or upgrade work on the *Zafir*/*Qahir* or *Scud*-B, which Egypt was said to be still pursuing in 1991. Egypt is also rumoured to have assisted Romania in developing a medium-range missile similar to the *Condor*/*Badr*.[177]

Infantry Weapons and Equipment

One of the earliest Egyptian manufacturing activities in the defence sector was the licensed production of Swedish submachine guns, starting in 1952. In later years copies of other rifles also appeared: the 7.92-mm *Hakim* (based on the Swedish 6.5-mm *Ljungman*), 7.62-mm *Rashid* (based on the Soviet SKS), and the Soviet *Kalashnikov* AK-47. Ammunition for these and other weapons was also produced in Egypt, and by 1969 local factories were manufacturing a copy of the Soviet RPG-7 anti-tank launcher and its sight (though not the projectile) under licence. Little other infantry equipment was made in Egypt, however, and none was designed there, with the exception of a modified version of the Soviet *Tokarev* 7.62-mm automatic pistol (altered to 9mm).

There has been minimal change in this situation since the 1970s. However, a major exception is the RPG-7, of which Saqr has produced a substantially modified version. The main improvement is in the projectile, which now has two types of warhead. One is *Cobra*, a HEAT round with an improved stand-off probe and fuze for shallow impact angles, using octogen explosives to achieve 500-mm penetration of steel plate at a range of 350m.[178] The second is *Haris* (*Home Guard*), which is an anti-personnel round with an impact radius of 60m and a range of 1,000–1,200m.[179] At peak rates, two shifts can produce 750 RPG-7 rounds daily.[180] Among Egyptian-made medium weapons are 60-mm and 82-mm mortars: the former is a copy of the Chinese Type-63, the latter a copy of the Soviet M1937 (but with a Thomson-Brandt baseplate).

The other area of infantry equipment in which the Egyptian

defence industry has made notable developments is that of support systems, from sights to range-finders and mine-detectors. Most of this work has been undertaken by joint venture or private sector companies. An example is the co-operation of Britain's United Scientific Holdings (USH) with the National Services Project Organisation of the Defence Industry Ministry. Starting in autumn 1985, they planned to produce general purpose 7×42 binoculars, SUSAT sights for the Enfield L85 *Endeavour* assault rifle, the AN/PVS-4 (NVS-700) night-vision device for rifles, and the LH 80 hand-held laser range-finder.[181] In April 1988, a joint American-Egyptian team announced the design of a new mine detector using a computer and radar device to locate objects up to 20m underground, which they were marketing at $10,000.[182]

Ammunition and Other Ordnance

Reflecting the experience of the destruction of the Egyptian Air Force on the ground by Israel in the early hours of the June 1967 war, among the air-launched products made or developed by Egyptian defence factories is a runway penetration bomb. First displayed by Qader Factory in late 1984, this 100-kg bomb is capable of penetrating concrete 400mm thick and making a crater 1.7m deep and 4m in diameter.[183] Adapted for use by a variety of Soviet and Western aircraft (MiG-17/21, Su-7/9, Tu-16, *Mirage-5*, F-4, it is released at an altitude of 120m and braked by a parachute for descent to 70m, at which point a rocket motor propels it on to target. The bomb became fully operational in 1989.

Guided Missiles

One of the aims of the AOI, upon its original establishment in 1975, was to develop a local capability for the manufacture of guided missiles and munitions. The first step in this direction was to obtain a licence in 1975 for the production of the British *Golfswing* ATGM. This project was not implemented, though, and the next step came with the formation of the Arab-British Dynamics Company, which obtained a licence for the production of the British *Swingfire* ATGM in 1977. Saqr also became involved in this in the early 1980s, working on modified versions of Soviet SA missiles.

Guiding Egyptian policy was the explicit aim of acquiring technical skills through licensed production or contract work for foreign suppliers. However, despite the acquisition of technical skills and experience, the Egyptian defence industry has not yet been able to move into independent design work on guided missiles. Lack of funding forms an additional constraint preventing local companies from undertaking licensed production of a wider variety of up-to-date systems. A result has been to limit the extent of technology transfer involved, and to confine much of Egyptian production to older missiles, thus reducing their marketability.

• *Swingfire*: despite the cessation of Arab participation in the AOI in 1979 and the latter's effective suspension, the contract signed between BAe and the Arab-British Dynamics Co. for the local production of the *Swingfire* ATGM continued unaffected. The original agreement was for 5,000 missiles, with deliveries to the Egyptian Army starting in 1979. By 1982 500 were being delivered annually, with production rising to 1,127 in 1986 and 1,136 in 1987.[184] A number of *Swingfire* systems were also exported, among them 200 to Sudan. Besides the individual, ground-launched model, ABDCo. also designed a four-launcher system mounted on the C-6 *Jeep* licence-produced by the Arab-American Vehicles Co. (a second *Jeep* carried 13 additional missiles for resupply).

A main advantage of the *Swingfire* programme is that the Egyptian contribution was allowed to reach 80 per cent.[185] It has also provided a testbed for local modification of the system, with development of an improved version starting in 1983. A year later a parallel development programme was started to increase missile reliability and double its shelf-life. As a result, system weight was reduced from 300 to 171 kg, with less ground gear and easier maintenance.[186] Warhead improvement took longer, though, with tests on the *Swingfire* warhead MK-2 being carried out in January 1988. Despite these improvements, *Swingfire* is essentially an obsolete system, thus explaining repeated Egyptian interest in obtaining licences to produce more modern weapons: concrete plans were first made for the *TOW* ATGM in 1982, followed more recently in 1989 by reference to the *Dragon*-2.[187] Failing so far to find funding or political approval for these new projects, ABDCo renewed its original *Swingfire* contract with BAe in January 1988 for an additional ten years, hoping to prolong the missile's operational life.[188]

• *Saqr-Eye*: along with the *Swingfire* ATGM, the *Saqr-Eye* programme has probably been one of the more successful Egyptian missile projects, in terms of entering serial production and providing technical experience. *Saqr-Eye* is an improved version of the Soviet-designed SAM-7, of which the Egyptian Army received large numbers before 1973. In combat, the SAM-7 proved to suffer from a number of drawbacks, including limited range and lethality, and high vulnerability to decoy measures.

The decision in principle to produce locally a man-portable, shoulder-fired, anti-aircraft missile was taken in 1982, with interest in a system similar to the American *Redeye*. This was part of a broader missile production programme being studied that included the American *TOW* ATGM and the French Matra *Magic* R-550 AAM. The aim was to obtain a production licence for an Egyptian company as prime contractor for the original designers. However, the estimated programme cost of $400m was too high, and the Egyptian choice finally fell on upgrading the SAM-7.[189] Modification of the latter commenced in 1984 at Saqr, with low-volume production starting in March 1985. By the time full-scale production was attained in early 1986, capacity stood at some 2,700 units annually.[190]

In order to modify the SAM-7, Saqr reverse-engineered its electronic guidance components.[191] It then added a Teledyne infra-red sensor, a Thomson-CSF IFF unit, and Rank Pullins night sight.[192] Range was increased to 4,400m, and operating altitude to 50–2,400m – an improvement over the original SAM-7 but still below the performance of the Soviet improved Model 1 version – while additional work was carried out to improve warhead lethality and increase ECM resistance. The result, *Saqr-Eye*, has been supplied to the Egyptian armed forces, while Iraq ordered a $50m consignment in 1988.[193] In addition to individual launchers, there is a vehicle-mounted version in which four or six tubes are mounted in one or twin banks at the back of a jeep or APC.

• *Sinai*-23 and *Nile*-23: other versions of the *Saqr-Eye* have also been developed. In response to a five-year modernisation plan issued by the Air Defence Force in July 1984, competition was opened for a mobile, low-level, defence system based on a gun-missile mix, to accompany ground forces in combat. Two proposals were made in late 1985 based on the twin-barrelled 23-mm AA gun and the *Saqr-Eye*: the *Nile*-23 and the *Sinai*-23 with the main difference between them being that the latter incorporated six rather than four *Saqr-Eye* launchers. Following test trials that started in April 1986, the *Sinai*-23 was chosen. A production deal was signed between the AOI and Electronique-Serge Dassault (ESD) in August 1987.

According to the Egyptian Air Defence Commander, General Sayyed Hamdi, Egyptian engineers and technicians provided extensive ideas and experience for the development of the *Sinai*-23 (and *Nile*-23).[194] But ESD was in charge of overall design, at the head of a team of European companies. As a system, *Sinai*-23 consists of a battery of five air-transportable M-113A2 vehicles: one carries an ESD-made RA-20S pulse doppler radar for detection of low-flying aircraft at over 11km; the other four each mount a twin 23-mm gun and six *Saqr-Eye* launchers on a turret.[195] ESD also provided the weapons-electronics interfacing, while Hispano-Suiza provided the turret. Furthermore, due to dissatisfaction with the performance of *Saqr-Eye*, Thomson-CSF/Brandt Armements was given responsibility for further improvements to bring it up to American *Stinger* standards. Egyptian officials claim that a large part of production will take place in Egypt (at Saqr and Abu Za'bal principally) for delivery from 1990 onwards, but in the absence of specific information it appears that local input is mainly confined to assembly and adaptation of the M-113s. An additional prospect is modification of *Sinai*-23 for operation on the *Fahd* APC.

• *Amun*: this is essentially the Swiss-Italian (Contraves-Italiana) *Skyguard* short-range, air defence system, consisting of two twin 35-mm cannons (Oerlikon) and two four-round *Sparrow* AIM7-F (Raytheon) SAM launchers, modified for Egyptian use. As in the case of *Sinai*-23, *Amun* is the result of design work undertaken by a foreign firm to Egyptian Air Defence specifications, based on a letter of intention signed in 1981. In 1983 a total requirement for 18 batteries was placed, worth $400m.[196] Following tests in Italy in October 1984, delivery to Egypt commenced in 1985 and was completed in 1986. Kuwait signed an order in June 1988 to buy a number of *Amun* systems for an undisclosed amount.[197] Unlike *Sinai-23*, however, *Amun* is not produced in any stage in Egypt.

What makes *Amun* partly an Egyptian product is that the modifications the original *Skyguard* reached up to 50 per cent of manufactured content, according to Defence Minister Abu Ghazalah.[198] Of these modifications, 16 were developed at Egyptian plants, which hold patent rights.[199] On this basis Egypt has insisted on receiving a four per cent share of all sales of *Amun*.[200]

• *Early Bird*: another attempt to prolong the life of air-defence equipment already in service is the Egyptian copy of the Soviet SAM-2, called *Early Bird*. Work started on this system in 1984, with the aim of extending its operational use for another ten years. As with the SAM-7 the guidance system was copied through reverse-engineering before going through modification. Originally, production was planned for 1985, but as in the case of several other systems, testing was still proceeding as late as March 1986, during which a hit probability of 98 per cent with three missiles was claimed.[201] Despite such a record, there was insufficient improvement over the original version: the new missile had the same basic dimensions, weight, and range performance. That the Egyptian defence industry could manufacture the SAM-2 – including such components as the two-stage motor with its solid boost propellant and a liquid fuel sustainer – was a definite advance, but this consideration was not enough to persuade the Air Defence Force to acquire *Early Bird*, as it considered the basic system too old.[202] Four years after programme initiation, some experience had been acquired but little else.

• *Other missiles*: in order to meet Air Force and Air Defence requirements, the Egyptian defence industry and armed forces workshops have undertaken a number of minor modification programs on SA and AA missiles in service. The most ambitious of these programmes is the partial production under licence of the French Matra *Magic* R-550 AAM. A second project was the establishment of a joint repair and assembly plant at Helwan for the Ford Aerospace AIM-9P-3 *Sidewinder* AAM, scheduled for the end of 1988.[203] Within the armed forces, workshops have undertaken maintenance of *Hawk, Crotale,* and *Chaparral* SAMs, on the basis of which the last two systems have received modifications, according to Abu Ghazalah.[204]

Electronics

Although underdeveloped, the electronics sector is nonetheless a significant one that is receiving increasing attention in Egypt. In 1989, for example, the AOI asked for bids to establish a new electronics factory in addition to its existing facility. Between them, the AOI and the NODP produce a variety of defence products under licence from Western companies. Principal among these are tactical communications equipment, which includes receivers and transceivers for military telephones (the SB-3614 automated switchboard, for instance), tactical radios (Watkins-Johnson wide-range high-frequency receiver), and tank radios PTR 1410E. More recently, Egyptian officials claimed in 1989 that the industry was developing a local military radio system.[205] Meanwhile, the optronics sector is developing, with a variety of basic optic equipment like binoculars and sights being produced within the private sector, though advanced uses of electro-optics and fibre-optics have not yet been developed. This activity takes place within the private sector, which is also mainly responsible for the modest amount of computer work done locally.

In addition to such tactical equipment, the defence industry is supposed to have designed and produced a complete system for automated command and control for use in Egyptian E-2C AWACS aircraft.[206] In the same strategic context, Benha Company has started production of the American-designed Westinghouse AN/TPS-63 early warning radar. Starting with a local contribution of 47 per cent that is meant to reach a level of 92 per cent by 1993, savings of hard currency are expected to rise from 30 per cent initially to 60 per cent by completion of the programme, with a net savings per unit produced of $1m.[207] Benha is also supposed to produce shelters and switchboards for Egypt's strategic communications network, which however has been designed by the American GTE Government Systems Corporation.[208]

Naval Equipment

This is the least developed branch in Egypt's defence industry, reflecting in part the predominance of air and land forces in the Egyptian military structure, and in part the general weakness of the country's naval industry. Yet the naval industry is the most developed among Arab countries, and more advanced than in several other arms-producing countries in the Third World.

After a modest beginning, a boost came in July 1969 with the signing of a Soviet-Egyptian agreement under which the USSR would provide training and expertise for ship-building. The Alexandria shipyard was, moreover, largely constructed under Soviet direction and equipped with Soviet machinery. At present, at least four shipyards are active in defence-related projects. These are Arab Contractors Shipyards (Alexandria), Timsah Shipyard (Ismailiyyah and Abu Kir), Castro (Port Said), and Canal Naval Construction (Port Fuad).[209] Mainly involved in civilian work, these establishments have undertaken some construction work for the Egyptian Navy, especially since 1980. An additional facility is the Central Technical Naval Centre, which was responsible for most of the production work on Egypt's *October* class FAC/Ms.

The earliest major naval construction programme was undertaken by Castro at the beginning of the 1960s. This consisted of three large patrol craft – *Nasr, Nimr*, and *Tha'r* – which were ordered in 1960 and launched in May 1963.[210] Each boat displaced 110 tons, mounted a single 20 mm gun, and was powered by 2 Maybach diesel engines, giving a total 3,000hp for a speed of 24 knots. They are still in service, with an additional two being completed in 1984.

A long period of inactivity followed the launching of the first three boats, during which the Egyptian Navy relied on Soviet and Chinese vessels. But after the experience of the 1973 war and the progressive breakdown in Egyptian-Soviet ties, Egypt started local production of a missile-armed, fast attack craft. This was the *October* class, which was a copy of the Soviet *Komar*. Six were built in Alexandria between 1975 and 1976, and equipped with Western armament and electronics suites. The boats were later refitted in Britain in 1979–81, and in 1984 Egyptian shipyards constructed another two of the class. *October* is powered by four CRM-18 D/S-2 diesels, giving 5,400hp for a speed of 40 knots (at 30 knots the range is 400 miles). Displacing 82 tons and carrying a crew of 20, the boat is armed with four twin-30 mm guns and two *Otomat* SSMs.

With the exception of four 10-ton patrol craft ordered from Canal Naval Construction in 1983, the next programme of the Egyptian shipyards was the *Timsah* class. In 1980 an order was placed with Timsah Shipyard for a series of six coastal patrol boats. The Arab Contractors Shipyard was to assemble six of the engines. The first three boats were delivered by December 1981, with another three a year later. A second order for six *Timsahs* was placed in summer 1983 and completed in 1984, following which a third batch of six was requested in January 1985. However, the latter were slightly modified, as the small funnel was removed to improve visibility, the two engines were replaced with MTU 12V 331 TC92 diesels (generating 980kW or 1,330hp each) to achieve speeds of 25 knots, and the guns replaced with twin Oerlikon GAM-B01s.[211] The original *Timsah* displaced 99 tons, had a length of 29m, and was powered by twin MTU 8V 331s.[212]

The next line of activity was the assembly of the American-designed *Swiftship* '93ft' class of patrol boats. Three were bought from the USA, and six were assembled locally and delivered from April 1985 onwards. This project coincided with the *Timsah* programme, as both *Swiftship* and the modified *Timsah* used a locally-assembled MTU 12V 331 TC92 diesel engine generating 1,332bhp. Two of these gave *Swiftship* a speed of 27 knots. The boat displaces 102 tons, has a range of 900 miles, and mounts two twin-30 mm guns.[213]

Having successfully met this initial demand, Egyptian shipyards are supposed to undertake further licence-production of *October* fast attack missile craft.[214] However, local naval construction capabilities remain somewhat limited, as demonstrated by the fact that upgrade work on the four Chinese-built *Romeo*-class submarines in Egypt's possession (to fire Harpoon anti-ship missiles and Mk37 torpedoes) was entrusted to Tacoma in the USA rather than to local yards.[215] Another option is to expand work through inter-Arab co-operation. Late in 1988 Egypt and Iraq agreed to overhaul 14 Iraqi merchant ships in Egyptian shipyards, indicating a potential for more ambitious joint projects.[216]

The Egyptian Defence Industry: the Obstacles

In assessing the current situation of the Egyptian defence industry, several obstacles stand out that hinder its development. The most obvious is the lack of funds, which has led frequently either to the cancellation of proposed projects or to repeated delays in their implementation. Towards the end of the 1980s, Egypt was estimated to need \$4–6bn in order to develop its arms industry, but was facing severe difficulty in obtaining loans to its high level of indebtedness and poor credit owing rating.[217] Nor are the prospects any better for the early 1990s: even after the write-off of \$14–16bn in civil and military debts during the 1990–91 Gulf crisis, the country still owes some \$30bn, of which at least \$3–6bn are for arms. (Annual debt servicing reached \$1bn in 1989, of which \$600m was to the USA.)

Financial constraints have prevented Egypt from providing seed or investment capital for new ventures, even where these have also been partly-financed by foreign loans or lines of credit. And, as the basis of Egyptian military industrialisation strategy since 1975 (and definitely since 1979) has been to acquire skills and know-how through assembly or production under licence, lack of funds has curtailed the entire process by limiting the number and variety of projects. A classic example of the consequences of financial constraints for the defence industry is the cancellation of the *Lynx* helicopter programme following the withdrawal of Arab support for the AOI in 1979. The financial problem is also evident as regards missiles, where lack of capital undermined an ambitious plan to produce the *TOW* ATGM, Matra *Magic* AAM, and the *Redeye*-type SAM in 1982, compelling Egyptian industry to drop the first, content itself with part-producing the second, and lastly copying the Soviet SAM-7 (instead of a more up-to-date Western weapon). Similarly, rather than produce the *Crotale* SAM as it had hoped in 1983, Egypt only overhauls it locally.

An associated problem of financial constraints is the adverse effect on the economics of local defence production. As mentioned above, the military industrialisation strategy is predicated on licence work, implying a need to increase the local share of production as much as possible in order to acquire more skills and production technology. Yet local industry can increase its work share only if the total production run is large enough. This in turn depends either on providing enough Egyptian capital to absorb the costs of raising local input or on ensuring large enough export markets to take up the surplus production. As an Egyptian defence industry official explains: 'At AOI we produce aircraft (the *Tucano*) with a local input of 72 per cent. Were the other Arab countries to unify their requirements [for a trainer] we could attain a local share of 100 per cent. Conversely, if the number [produced] was a lot less, then our input would drop too.'[218]

This not only affects Egypt's industrialisation strategy, however; it also affects its exports. One of the country's main advantages is lower production costs – mainly for labour – and so a greater local work share means lower end-product prices, thus boosting marketability and exports. Conversely, inability to increase the local work share raises end-product prices, making them less competitive. Thus Egypt sold *Tucano* trainers to Iraq at the same price as that offered by Brazil, thus eliminating its main commercial advantage.[219] This problem becomes even more pressing when it comes to major, modern weapon systems. These require an even greater initial investment by Egypt that it is poorly-placed to provide. Yet it is systems such as these – *Mirage*-2000, F-16, and M-1 – that are most desired by major Arab and Third World importers, that return the largest profits, and that offer the greatest technological benefits for Egyptian defence industry. But under present circumstances it is highly likely that unit prices for such items produced under licence in Egypt would be higher than for the originals – unless major export orders had already been made, as with the *Tucano* programme.

The preceding suggests that the problem does not lie with Egypt alone. The willingness of Arab states especially to purchase Egyptian products is a decisive factor in the economics of its defence industry. Indeed, former Defence Minister Abu Ghazalah singled out this problem for particular attention. In August 1986 he complained: 'We have good armaments production, but the Arab countries do not buy from us. Saudi Arabia, for example, bought *Hawk* planes from Britain. We have the *Alpha Jet*, which is a better plane. Saudi Arabia bought the [*Tornado*], and we have a factory capable of producing the *Mirage*-2000. There are political factors that control this.'[220]

Even after the full restoration of Egyptian-Arab ties in 1988–89, a number of Arab states still opted to buy systems from Western suppliers, although the same items or similar ones were on offer in Egypt. Within the space of one year, 1988–89, Saudi Arabia chose the Bell 406CS helicopter over the Aerospatiale 342L (assembled in Egypt), the UAE bought French-made *Mirage*-2000s (rather than support the Egyptian assembly programme of the same aircraft), Oman ordered the *Hawk* trainer/ground support aircraft (rather than Egyptian-assembled *Alpha Jets*), and Kuwait opted for British-made *Tucanos* (rather than Egyptian ones). Furthermore, although Egypt had only recently confirmed its deal with the USA to produce the M-1 tank locally (at significant expense), Saudi Arabia requested 315 vehicles direct from the USA in October 1989, for a contract value of $1.5bn (half of which was for follow-on support).[221] The only exception to this pattern was the possibility of Egyptian participation in providing 50–60 *Alpha Jets* and 50–60 *Mirage*-2000s (possibly rising to 120–150 of each type eventually) to Iraq as part of the Iraqi-French deal signed in June 1989.[222] However, even in this case there were potential problems for Egypt, as Iraq was also negotiating with both Brazil and France for local production rights for the *Tucano* and the *Alpha Jet*.

The political reasons for Arab reluctance to purchase Egyptian products or to invest in the arms industry are not one-sided, however. According to one report, the GCC states were reluctant to revive their industrial co-operation with Egypt and rejoin the AOI in the late 1980s partly because of the closeness of Egyptian-American ties, and partly because of Egypt's refusal to cancel certain military and industrial programmes.[223] The Gulf states also felt slighted by the Egyptian-American MoU issued in 1988, as it dealt with their security without consulting them or involving them in formulating the draft. This situation indicates the sort of problem that belies Egypt's hopes to rebuild the AOI as a medium for inter-Arab co-operation, in which 'the money would be Arab, but control and management Egyptian'. Post-Gulf crisis trends, while reversing the previous order of triangular American-Egyptian-Gulf relations, scarcely point to a deepening of military and industrial ties between Egypt and the GCC states.

Other problems confronting Egyptian defence exports are technical, and relate to expected support and other services (such as training) from the supplier. Financial factors are naturally important: Egyptian contracts offer little net saving over those with the original suppliers, despite repeated claims by senior officials that local products are substantially cheaper. Indeed, Qatar annulled a £4m contract for *Fahd* APCs and AD weapons in 1988, claiming that they were over-priced.[224] Loss of competitive edge is partly owing to commercial and administrative practices within the Egyptian defence industry and state bureaucracy. Specifically, foreign observers suggest that the amount of paperwork involved in finalising deals, and the amount of commission to be paid to middlemen, render the total cost to the customer on a par with that offered elsewhere.

The same twin problems have hindered a greater private sector role in military production. Red tape and commissions have deterred local businessmen and industrialists, often short of resources, from investing in defence ventures. This has, moreover, affected the growth of upstream industries associated with defence production, such as steel and alloy plants. Thus two major programmes for the local production of carbon steel, titanium oxide, and other alloys in Egypt were floundering in 1989, because, although foreign capital could be found, not enough private Egyptian capital was forthcoming to fill the 51 per cent requirement (that this proportion of any joint venture should be Egyptian-held).[225]

Negligible private sector participation in the defence industry and associated branches has had a significant debilitating effect (especially as, despite official founding statements, the AOI has attracted virtually no private Egyptian capital, as distinct from European and American investment). It has weakened accountability and therefore distorted commercial thinking and assessments of feasibility. The industry remains overwhelmingly a state one, with substantial room within it for personal aggrandisement, as the experience under Ashraf Marwan's chairmanship of the AOI suggests.

In view of the preceding remarks, the future development of the industry remains bound by a number of interlocking problems that limit both the technological and the commercial benefits it can ever hope to achieve in the absence of radical changes in its structure and methods of operation. In practical terms this means that the traits characterising its efforts to date will persist. These are: the predominance of foreign assistance in all levels of design and production; weakness of specialised areas, such as electronics and metallurgy; limited local input; and limited savings on production costs and thus reduced export competitiveness.

These traits are evident in the review of defence industry activities. In all but a small number of projects, foreign assistance was essential to the success of the local production. Thus even such items as the *Amun* air-defence system, the *Saqr-Eye* SAM, and the *Saqr*-30 MRLS – all billed as Egyptian designs – were, in fact, designed either wholly or partly (in crucial components) by foreign experts. Quality control experts on major production lines such as the *Alpha Jet* are non-Egyptian, although this is an important area of technology transfer. Naturally, some of the shortcomings of local industry – such as the underdevelopment of electronics – are related to the general modesty of Egypt's scientific base. But here a concerted effort becomes even more necessary, and is more lacking.

Indeed, a general problem is the weakness of R&D, with little participation by universities and similar establishments outside the actual defence companies, and with few independent, dedicated R&D centres. Combined with negligible private sector participation, the weakness of R&D and the 'science culture' place a daunting ceiling on the ability of local industry to undertake ambitious projects or achieve technology transfer.

Final Comment

The various obstacles and problems notwithstanding, Egypt's defence industry remains the leading enterprise of its kind in the Arab region. It is the best equipped to sustain and expand its activities, moreover, both quantitatively and qualitatively. This is largely owing to its substantial industrial experience and scientific heritage, especially since the early 1950s.

Egypt's advantages are reflected, among many other things, in its establishing of technical training centres within the armed forces, which prepare personnel for the repair and maintenance of military equipment. The Armed Forces Technical Institute (AFTI) in Cairo, for example, trains officers for all four services, while the Air Force's Technical Training Institute (TTI) additionally produces 500 NCOs annually after a three-year course.[226] This general capability assists the arms industry in offering after-sales services and training to potential export customers. To the latter, General Sayyed adds that Egypt imposes no political commitments or conditions on its customers.[227] The reverse side of this advantage is that Egypt is equally well placed, at least in theory, to enjoy and promote inter-Arab military industrial co-operation.

Indeed, an added cause of Egypt's potential strength in the inter-Arab context is the size, experience, and manpower of its civilian manufacturing industry. Notably, in the early 1980s, the arms industry accounted for 8 per cent of total manufacturing employment, or 15–20 per cent of employment in the 'modern' manufacturing sector.[228] The private sector is also relatively substantial and active, with an official policy since the 1970s of promoting industrial liberalisation and related financial incentives. These elements form the basis for vitality and viability, and contrast favourably with the situation in other Arab countries (including those, like Saudi Arabia, that encourage the private sector but lack real industrial ability).

However, the role of both the civilian feeder industry generally and of the private sector is undermined by the bureaucratic and financial problems mentioned earlier. This has resulted in the regression of the local automotive industry, for example, which is not only operating at one-half to one-third of capacity but is also reverting to the assembly of cars, after earlier reaching the stage of part-producing them, owing to lack of hard currency.[229] The problems of the Egyptian economy as a whole, as well as of indebtedness and difficulties in repayment, mean that local industry has been even less able to achieve offset or barter deals with outside suppliers of parts or designs, which add to the financial and technological obstacles.

In conclusion, the Egyptian defence industry is sufficiently advanced and remunerative to maintain itself at the current level, but it requires a drastic change in certain areas before it can achieve any commercial or technical breakthrough. On the surface, the main need is for capital. Thus Egyptian officials have eagerly raised the prospect of a return to Egyptian-Arab co-operation within the AOI. But they have also made clear that this does not mean a return to the old formula, which left the industry vulnerable to sudden changes of political mood as in 1979. The working motto at the end of the 1980s was 'Arab capital but Egyptian direction', implying sole Egyptian control over assets and decision-making. However, this is unconvincing to Arab financiers, whether governmental or private, not least owing to distrust of the bureaucratic system and profiteering. Egyptian thinking, meanwhile, contains an element of self-delusion that is likely to undermine commercial and technical planning.[230]

One alternative that has been suggested is to attract increased foreign investment, or to operate on a sub-contract basis for Western arms companies. Arguing in favour of such an approach, Egyptian officials point to the fact that skilled labour costs are low (one-sixth American levels) and trade unions inactive, and that infrastructure and management can be provided locally.[231] The technical and financial rewards of this are likely to be limited, though, and Western multinationals have been generally reluctant to invest in Third World defence industries. Another, more realistic option therefore is to increase exports in order to obtain needed funds. But to do so Egyptian defence products must become more competitive, both technically and financially. This requires the reduction of production costs – through improved management and marketing, or the generation of greater private participation – addressing user needs more carefully, developing local R&D, and improving follow-on services.

In each and every case, the crucial prerequisite for any advance is to achieve internal reforms and reorganisation within the defence industry in particular, and within associated feeder industries and the financial sector. It is such measures that can allow an increase in funding and in local content, and therefore cut down overall costs and export prices, or at least gain in quality. In the absence of major restructuring, however, the Egyptian defence industry is unlikely either to be given a free hand with Arab funds or to generate sufficient resources to keep pace autonomously with military hardware and technological developments.

5

Iraq: Emergent Arms Producer and Regional Power

Introduction

Iraq emerged relatively suddenly, at the end of the 1980s, as a significant arms producer in the Arab region. Almost overnight, it seemed, it had become the second major Arab military industrial country, with ambitious plans for future growth and the funds to implement them. Besides the immediate significance of this development for the course of the Iraq-Iran war and for the volume and nature of Iraqi arms imports, the local defence industry became an added pillar of Iraqi regional power and 'reach'. The massive destruction wreaked on Iraq during this and the 1991 Gulf war crippled the country, however, as much by imposing astronomical bills for debts, reparations and reconstruction and by inviting punitive sanctions on military technical transfer, as by direct damage to factories, depots and R&D centres.

The first indication of the existence of a significant indigenous defence production capability in Iraq came with the employment of chemical weapons against Iranian troops in 1984. The next major indication was the firing of some 300 surface-to-surface missiles against Iran in spring 1988, most of which had been modified or produced in Iraq. Only following the end of the Gulf war in August 1988 did the Iraqi industry start to exhibit the systems it had developed during the preceding years.

The country's first attempts to establish a local defence industry came in the mid-1970s, but it was only midway through its gruelling war with Iran, in 1984, that a major new development and expansion programme was launched. Since then, in the space of a few years, Iraqi factories have come to produce a wide variety of infantry weapons, artillery systems, ammunition, air-to-ground munitions, and electronics components, some of which have been used to modify tanks and aircraft in active service with the Iraqi armed forces. With a reported 100,000 employees at the end of the decade, this sector was equal in size to that of Egypt.[1]

Aims of Military Industrialisation

Iraqi military industrialisation is especially striking because it is the first instance of an Arab country building up its indigenous productive capability during war. In the main, this took place in the brief period between 1984 and 1990 which indicates a high rate of financial and human investment and suggests the existence of a scientific and technical base sufficient to allow such rapid expansion. Owing to the impact of the war, furthermore, the items produced have been selected entirely according to immediate combat needs, and not to commercial or other factors, such as bureaucratic syndrome or political and regional prestige.

Nonetheless, Iraqi officials have resorted to classical arguments in order to explain the rationale for establishing a major defence industry in the country. One summed up Iraqi thinking by focusing on the following economic considerations:[2] (1) feed the civilian sector with the results of military R&D; (2) absorb and utilise surplus labour; (3) activate associated industries, such as steel and chemicals production; (4) improve the national balance of payments and save hard currency; (5) increase the share of locally-added value and so improve GDP; (6) reduce unit costs by expanding production runs; (7) achieve large financial returns from exports through the high profit margin involved in the sale of military products in world markets; (8) accumulate technical experience and expertise; and (9) use national resources – human, financial, and natural – better and so reduce dependence on the outside and vulnerability to world fluctuations.

The preceding remarks reflect little more than standard thinking on the issue, much of which has been open to question over the past decade. Far too little is known about the Iraqi defence industry and its economics to judge whether the considerations listed above were realistic in any sense, even before the catastrophic second Gulf war, or if they could have been borne out in practice. For instance, Iraqi civilian industry was still seriously underdeveloped, and was unlikely to benefit from military R&D, though this was a possibility in the areas of steel, chemicals, and electronics. Moreover, the extent and nature of local R&D is another major unknown, and, given the concentration of all resources for the war effort against Iran until mid-1988, the issue of military-civilian interaction is highly theoretical. Involvement of the Military Industrialisation Authority itself in civil construction – of bridges, dams, and grain silos – does not change this aspect.

The amount of accumulation of expertise in defence production also cannot be gauged, as there is little indication of the proportion between Iraqi and foreign input into design and manufacture. The fact that at least the earlier military factories established in Iraq were handed over on a turnkey basis does not suggest a high local input, nor a great opportunity for transfer of expertise. However, statements in the late 1980s that Iraq had sought to import programmable machine tools and produced some of its own suggest an attempt to shift from assembly to manufacture, and thus the elements of an independent technological capability.[3]

Similarly, it is impossible to assess the economic impact of local defence production on the Iraqi balance of payments or GDP up to 1990. At best, an educated guess can be hazarded that, as the main Iraqi effort until then had focused on producing light and medium arms and ammunition – consumption of which was enormous in the earlier war – or on modifying existing systems in service and not on producing major complex systems, such as tanks and aircraft, the total cost of local production would not have exceeded by a wide margin the import price for an equivalent amount of *matériel*. Of course, few savings of hard currency would have been made at that initial stage, owing to the need for outlays to purchase manufacturing equipment and technical services from foreign companies.

Conversely, the strategic benefits for Iraq were far more obvious and convincing. As the same commentator noted earlier, producing needed defence items locally tends to: (1) circumvent control or interference from supplier countries; (2) avoid vulnerability to arms embargoes; (3) pre-empt political pressure on suppliers from third parties; and (4) ensure the secrecy and security of local defence needs and projects. The Iraqi experience is evidently embodied in these remarks. On the first count, Soviet reluctance to supply Iraq with certain weapon systems – or even to suspend shipments altogether – posed a constant strain on bilateral relations throughout the first Gulf war. American pressure on France, and international pressure on the USSR in 1987–88 also threatened to hinder needed arms supplies. But probably the most compelling reason for the Iraqi decision to expand its defence industry was the move by France and other suppliers in 1984 to suspend military transfers until outstanding debts had been paid.[4]

Pre 1990 Status

During the eight years of the first Gulf war, Iraq is estimated to have imported $32bn worth of military equipment.[5] In the post-war period, it needed both to maintain a high state of combat readiness and to constantly modernise its armed forces. Given the sheer volume of previous military expenditure, the decision to provide as many needs as possible locally seemed sound enough. Possessing the world's second largest reserves of oil, Iraq, moreover, appeared capable of mobilising extensive financial resources, despite its massive war debt (estimated at $75–80bn) and needs for reconstruction.[6] Indeed, it was reported in December 1988 that the Iraqi government was planning to invest $20bn in local military industrialisation over the coming years.[7]

As part of its ambitious plans before the 1990 Gulf crisis, Iraq was seeking to establish its own aerospace industry. The first practical step towards this goal was taken in June 1989 with the signing of an agreement with France for the purchase of a number of combat and trainer aircraft. Worth FFr22–25bn ($3.2–3.5bn), the deal involved licence-assembly and partial production of 50–60 *Mirage*-2000s and 50–60 *Alpha Jets* (possibly rising to 120–150 of each type).[8] The possibility was mooted of co-operation with Egypt, which already assembled or produced parts for both aircraft, although this remained unconfirmed. Meanwhile, Iraq was gradually producing a growing number of sub-systems for the Soviet T-72M1 tank, with the aim of eventually producing the whole tank under licence.[9] With this venture, Iraq would have extended its military industrialisation efforts into most of the main areas of defence production, promising to put the country on a par with Egypt.

Organisation of the Defence Industry

The Iraqi defence industry is completely state-owned. For a decade it was directed by two government bodies: the Military Industrialisation Authority (MIA) and the General Organisation for Technical Industries. The former was effectively an autonomous organisation that supervised the early ordnance factories set up in the second half of the 1970s, while the latter oversaw advanced technological projects (such as chemical and missile research in the 1980s) under the direction of the Ministry of Industry. Only in 1987 were both bodies brought together in the newly formed Ministry of Industry and Military Industry, which subsumed the previous Ministry of Industry.[10] However, the MIA still took practical responsibility for the running of the industry, by co-ordinating the activities of individual branches and factories within a central staff, while remaining under the overall supervision of the ministry. At its formation the new ministry was headed by Brigadier Husein Kamel, with Lieutenant General Amer Hammoudi al-Sa'di acting as first deputy (and becoming minister in spring 1991). An additional body involved in the development of the defence industry is the Scientific and Development Committee headed by General Ammar, which undertook modifications to equipment in service and standards specification work.

The Iraqi defence industry comprises a number of separate productive and research establishments, which employed a total of 100,000 persons at the end of the 1980s. First to be set up were powder and propellant factories supplied by the USSR on a turn-key basis in 1976–78.[11] A small arms factory was also established, producing Italian automatic pistols under licence from 1976–78 onwards.[12] In 1980, an important new production line was started. This was the Sa'd-13 factory near Mosul, which was set up on a turn-key basis by Thomson-CSF to produce electronic components.[13] Little else is known definitively, but in 1985–86 Iraq obtained equipment for the establishment of a new plant. This, a complex known as Sa'd-16, was also based near Mosul and was apparently responsible for the development and production of the heavy SSMs used by Iraq in the latter stages of the war with Iran.

It is not fully known where production of other military items was undertaken, but until 1987 the Military Industrialisation Authority was in charge of the small arms, ammunition, and mortar factories (among others). Furthermore, it is not known if some systems such as the modified Ilyushin-76 EW aircraft and the T-55 upgrade and T-72 assembly were produced at Air Force and Army workshops, rather than in dedicated factories. Indeed, the role of armed forces workshops within the defence industry is unclear, though some participation is to be expected given the experience of other countries, such as Egypt and Israel, and given the extensive experience army technicians must have gained during the war with Iran in repairing, maintaining, and adapting equipment.

Another significant element in the organisation of the Iraqi defence industry is the role of civilian ministries and enterprises, and of special commercial procurement agencies operating abroad.* Examples of the former are the Nasr State Enterprise for Mechanical Industries and the State Establishment for Pesticide Production: the first was involved in the ballistic missile programme and in other defence ventures, while the second acquired the chemical precursors needed for chemical weapons production. Prominent examples of the Iraqi purchasing network abroad, according to Western sources, were the British-based Technology Development Group and Meed International. Together, these bodies were responsible for overseeing

* A summary of the procurement network is in *International Defense Review*, June 1991, pp. 567–70.

R&D and production in certain areas of the defence industry and for acquiring in a discreet manner the materials, technology, experts and other inputs needed.

Products and Activities

The main areas of Iraqi defence production are the following: infantry weapons; artillery; armoured vehicles and modifications; air-launched munitions; electronics; and aircraft modifications.

Infantry Weapons

As mentioned earlier, production of small arms was one of the first activities of the Iraqi defence industry. Since the mid-1970s, a wide variety of such weapons have been produced locally, along with a number of light or medium crew-served weapons. These have been manufactured in large quantities since 1980 to meet combat needs, along with a range of ammunition.

- *Small arms*: the first small arms to be produced under licence in Iraq were the Beretta 9-mm and 7.65-mm pistols, starting in 1976. These were followed a year later by the Yugoslav M72B1 light machine gun (7.62mm×39). It is not clear what proportion of the components were made locally, but Iraqi officials insist that all subsequent weapons were manufactured wholly in the country. In 1980 an Iraqi copy of the Soviet *Dragunov* sniper rifle (7.62mm×54) was designed, renamed *Qadisiyyah* and mounting a Heckler & Koch telescopic sight and cleaning kit. Later, Yugoslav assistance allowed local production of the M70B1/B2 assault rifle (a copy of the Soviet AK 47).[14] Along with these weapons, Iraq also produced hand grenades, rifle grenades, and a range of ammunition, the latter including Western- and Soviet-type 5.45-mm rounds for assault rifles.

- *Crew-served weapons*: Iraq produces anti-tank rocket launchers and mortars. The former have not been displayed and so their exact type is not known, but given the vast prevalence of the Soviet-designed RPG-7 it is probably the weapon under manufacture. The question then remains whether this was achieved with Soviet or possibly Egyptian help, and if the projectile as well as the launcher is produced locally. Also in production is the Soviet-designed AGS-17 automatic grenade launcher. Iraqi mortars are of 60-mm and 82-mm calibre, although the 120mm and the 160mm are apparently also produced in the heavier range. According to Iraqi official sources, production of mortars commenced in 1987 and reached over 4,600 almost immediately.[15] Iraqi factories also manufacture mortar shells in several calibres and types.

• *Other equipment*: among the *matériel* produced locally are anti-personnel and anti-tank land mines, of which Iraq used vast quantities in its war with Iran. Another item is protective gear and clothing against chemical attack, in the manufacture of which Egypt has co-operated with Iraq.[16] No information is available on Iraqi activity to produce optical gear (such as binoculars or rangefinders) or communications equipment, despite the dedication of the Sa'd-13 factory to electronics, with the help of Thomson-CSF.

Artillery

Artillery played a major role on both sides during the Iran-Iraq war: in support of massive infantry or armour attacks, in static warfare and attrition, and in the shelling of vital installations. Iraq in particular relied on massed artillery defensively, to create concentrated fire zones and pre-designated killing grounds into which Iranian infantry attackers were channelled. The war therefore reconfirmed the standard lessons that this arm is a principal cause of casualties, and a main reason for the extraordinarily high consumption rates of *matériel* in modern combat. In order to meet its pressing need for more artillery weapons and for ammunition, and to avoid the effect of embargoes or restricted foreign supplies in this category especially, Iraq pushed through several local production programmes. A secondary aim was also to develop or modify systems that were more appropriate to Iraqi needs and battlefield conditions in terms of terrain and manpower. This affected the development of special munitions as well as of armament.

• *Gun artillery*: two gun systems known to be produced in Iraq are the Soviet-designed 122-mm D-30 howitzer and a 155-mm hybrid. Although the former is also produced in Egypt, Iraqi manufacture has not been due to Egyptian assistance nor to reverse-engineering, but to Yugoslav technological assistance. Named *Saddam*, its specification indicates a weight of 3,335 kg in firing position and 3,440 kg in travelling order – greater than the original by 185 kg and 230 kg, respectively. This increase is partly due to the addition of hydraulic jacks to speed up time into and out of action. Maximum range has also been increased from 15.4 km (without RAP) to 17.1km. The main Iraqi innovation was to develop a new 122 mm round for use with the D-30, which is a bomblet dispenser.[17] The second gun is, in fact, a conversion of the Soviet M-46 130-mm gun, using the same trail and carriage but replacing the original barrel with a locally-produced, 45-calibre 155-mm one, for which ammunition is also made.

• *Mortars*: besides the lighter and medium mortars mentioned earlier (60mm and 82mm), Iraqi ordnance factories also manufacture the 120-mm and 160-mm calibres, as well as their shells. The former type can fire a 12-kg shell up to 10km, or a 16-kg shell to a range of 8.2km – the shell can use up to 10 increments.[18] It has also been mounted in self-propelled versions, though the 160-mm mortar has been shown only as a towed weapon.

- *Self-propelled artillery*: there are four types of self-propelled artillery system in development or production by Iraq, two mounting guns and two mortar-carriers.

The first self-propelled gun under development is the *Majnoon*, a 155-mm system that is essentially a copy of the South African G6, of which Iraq imported a number during the war with Iran. The principal difference is that the *Majnoon* is mounted on an articulated 6×6 vehicle (rather than a single chassis) in which the driver's compartment at the front is coupled to the fighting compartment by a pivot and two hydraulic arms and drive shaft. This vehicle, both by being articulated and wheeled, is especially suited to Iraqi terrain, particularly in the south with its winter mud and marshes. It is powered by a 450-kw Mercedes diesel engine and may have been constructed in Spain. The system weighs 43 tons and is armed with an 8.06-m barrel – produced in Iraq, probably from French designs – that fires a 45.5-kg ERFB-BB shell to a maximum distance of 38km, for a maximum rate of fire of four rounds per minute (sustained rate is one).[19]

For similar reasons, the second gun system under development with the *Majnoon* since 1987 – the *Fao* – is also based on the same articulated 6×6 vehicle, for a weight of 48 tons. But it mounts a 210-mm gun, which is not in service in any other country. Moreover, the *Fao* fires a 109.4-kg ERFB-BB shell to an unprecedented range of 57.3km. Rates of fire are the same as for the *Majnoon*, and both vehicles achieve road speeds of 90km/h and cross-country speeds of 60–70km/h.[20] Both weapons have been developed with European assistance, and especially French (for the barrels). A coastal defence version of *Fao* was under development before 1990 for use with radar fire control.[21]

Third is the self-propelled 120-mm mortar. A prototype was first displayed in April 1987, based on the Soviet MT-LB multi-purpose tracked vehicles. A standard 120-mm mortar was fitted in the rear compartment, firing out of the roof hatch. The barrel is supported by its normal bipod but without the baseplate, thus limiting its arc of fire. The second 120-mm carrier is also based on the MT-LB, and was first exhibited in spring 1989.[22] This system consists of a bank of four extra-long, 120-mm mortar tubes carried at the rear of the vehicle. A hydraulic lift system moves the weapon through 90° from above the vehicle to the ground, where two outriggers are fixed, for firing. The tubes can then move through an elevation of 40 to 80° and a total traverse of 20°, and can be fired singly or jointly. The vehicle carries 104 rounds and a five-man crew, who can take the system into action in two minutes and out of it in one.[23]

• *Rocket artillery*: Iraq produces a variety of artillery rockets for its Soviet-designed multiple launchers, especially those of 122-mm calibre. More significant is local design and production of other systems ranging from 81mm to 400mm. At the lower end of the scale is *Sajeel*, a 40-tube MRLS firing 81-mm rockets. It consists of two units, each with four banks of five tubes, mounted on a truck. A second local system is *Nasr*, which contains two four-tube banks of 240-mm rockets mounted on a truck using front and rear outriggers during firing for stability. Most impressive, however, is the *Ababeel* MRLS. This has been displayed in two versions: an older one mounting a large 12-tube launcher, and a more recent one with a single four-tube box. Capable of a 100-km range, the missile can carry varying payloads such as 300 anti-tank/anti-personnel (fragmentation) bomblets or 25 anti-tank mines.[24] The latter weigh 800 gm each and are activated by a magnetic fuze to detonate if a vehicle passes within 2m. Iraq may have obtained technological assistance from Brazil in developing some of these systems, as it has purchased *Astros*-II, SS-30, and SS-60 MRLSs from the latter and may have been involved in funding the SS-300 programme. There also appears to be a *Saqr*-30 system capable of firing chemical warheads; UN inspectors found a 122mm MRLS using binary chemicals.[25]

Tactical Ballistic Missiles

The appearance in combat of medium-range ballistic missiles in spring 1988 was one of the first major indications of the capabilities of the Iraqi defence industry. Most of the 300 missiles fired by Iraq at Iranian targets were of the *Hussein* type. This is essentially a modified Soviet *Scud*-B, of which the Army had some 20 launchers. However, the modifications introduced were substantial, involving an increase of range to 650 km, double the original. This may have involved halving the warhead weight, but, because it also meant shifting the missiles's centre of gravity, the modification required advanced technological work. This was apparently provided, at least in part, by the German MBB company, though Egypt may also have provided technical assistance.[26] Among other things, modification required the cannibalising of three old *Scud*-Bs to make every two *Hussein* missiles.[27] Iraq stated, after the second Gulf war, that it had 30 chemical warheads for *Hussein* missiles.[28]

The second medium-range ballistic missile to appear, though probably too late to be used in the first Gulf war, was the *Abbas*. This was a further modification of the *Scud*-B, for a greater range of 800–900km (depending on the source quoted).[29] Of equal importance is that *Abbas* was suggested to have a reported Circular Error probability (CEP) of 300 metres, compared to the 500 metres CEP of *Hussein*.[30] However, the apparent inaccuracy of *Scud* derivatives during the Gulf war of 1991 revealed either that their CEP was in fact much larger, or that Iraqi units had difficulty in aiming them correctly, owing in part to lack of intelligence information. It is possible that both versions may be capable of carrying the cluster warhead developed locally for the *FROG*-7s in Iraqi service, though this has not been confirmed.[31] Like *Hussein*, *Abbas* is carried on a launcher/erector transport trailer towed by a Scania truck known locally as *Waleed* (there is some confusion over whether the transporter or a different missile is called *Waleed*). Some reports suggest that the Iraqis modified imported heavy trailers, possibly of West German origin, to transport Scud missiles.[32] Besides the *FROG*-7 warhead improvement programme, Iraqi workshops have also modified the Soviet-supplied SAM-3 for the ground-to-ground role under the name of *Baraq*, though details of its payload and range are unknown.[33]

Before the 1990 Gulf crisis, Iraqi industry was reported to be working on a third missile with even greater range, called *Waleed*. This may have involved co-operation with Brazil, as the two countries had discussed the possibility of jointly developing a military satellite.[34] Another possible connection with the third Iraqi missile project is that Iraq may have funded the Brazilian *Avibras* SS-300 system, in return for technical assistance.[35] However, Western reports indicated that *Waleed* may in fact have been the Iraqi version of the Argentinian-Egyptian joint *Condor*-2/*Badr*-2000. The same sources suggested that, following suspension of Argentinian and Egyptian efforts under severe American pressure, Iraq may have developed the missile independently (albeit with the alleged help of Egyptian scientists), but also claimed that there had been serious technical difficulties resulting in flight deviations of up to 20 miles.[36] Egypt itself was reported to be continuing work on a *Said*-B derivative in 1991.[37]

Further indications of the scope and ambition of the Iraqi missile programme came at the end of 1989, when Iraq declared that it had an active space programme. This was followed, on 7 December, by the revelation that it had test-fired the booster rocket-stage of a missile capable of launching satellites into orbit.[38] This was *Tammuz*, with a reported range of 3,000 km. The USA confirmed the launch of a missile the next day, but could not confirm Iraqi claims that a second type had also been tested, namely the *Abed*, with a range of 2,000 Km. Later reports claimed that *Abed* had been developed with Argentinian co-operation, suggesting that it might in fact have been the Iraqi version of *Condor*-2.[39] None of these various missiles appeared during the Gulf war of 1991, nor were mentioned in lists of the Iraqi missile inventory. Though not a ballistic missile as such, a particularly telling example of Iraqi ambition and thinking is the development of the long-range *Ababil* ASM. Quite different from the conventional MRL carrying the same name, the mock-up displayed in 1989 of what may have been an air-breathing cruise-type missile revealed several notable features. Lacking the computerised and satellite-aided navigational abilities of cruise missiles, *Ababil* was nonetheless thought to have been designed to have inertial or mid-course command guidance and visual or infra-red terminal guidance. With a length of 6 metres and weighing 1000 kgs, it was supposed to have a range of 500 km.

Several plants were responsible for missile development and production before 1990. The major one was the Saad-16 complex, near Mosul, reportedly costing $200m to establish.[40] Among other facilities, it had a wind tunnel built by Austrian and German firms.[41] Elsewhere, the Masr State Enterprise for Mechanical Industries ran a plant at Taji, used Western-supplied machines and engineers to produce internal components and metal outer casings for the conversion of *Scud*-B Missiles into the *Hussein* derivative.[42] Nasr Enterprise also contracted British companies to provide computer software for the injection moulding of plastics for use in the missile programme.[43]

The 'Super-Gun'

In April 1990, British Customs agents seized a shipment of steel tube sections purported to be parts of a massive cannon with a 1,000 mm calibre. Over the next few days, Greece and Turkey announced the discovery of further tubes *en route* for Iraq, as well as components presumed to be loading mechanisms and breech-blocks. Other parts were reported from Italy, Sweden, Switzerland, Spain, and West Germany. Investigations revealed that British firms had already supplied Iraq with 44 tube sections, and previously exported other tubes with a diameter of 350mm and hydraulic mechanisms.[44]

Several hypotheses were suggested to explain the exact nature of the Iraqi 'super-gun' programme, but the essential conclusion was that Iraq was involved in developing a large calibre system capable of firing massive artillery projectiles or missiles to great distances. These plans stemmed from an Iraqi contract in 1988 with the Belgian-based Space Research Corporation, headed by the Canadian Dr Gerald Bull, to develop self-propelled guns of 155-mm and 210-mm calibre and enhanced ammunition, to run ballistics training courses for Iraqi personnel, and to conduct research on the super-gun project codenamed 'Project Babylon'.[45] The self-propelled guns were probably the *Fao* and *Majnoon* prototypes, but the super-gun appears to have been a development of the experimental HARP project run by Bull for the US Army in the 1960s. Reports in the Western press pointed to a variety of possible designs, including a 600-mm gun with a 800-km range and a 350-mm gun with a 160-km range – but the net conclusion was that the super-gun was barely in its early experimentation and development stage.[46] UN inspectors subsequently uncovered a 350 mm calibre super-gun in position for test-firing, at Jabal Hamrayn, 200 km north of Baghdad, and the unassembled parts of two more 350-mm guns and two 1000 mm guns at Iskandiriyyah, 50 km north of the capital.[47] The assembled gun had been test-fired once, to a range of 200 km, using specially developed gunpowder.[48]

Armoured Vehicles

During the 1973 Arab-Israeli war, Iraq's most prominent contribution was the despatch of two armoured divisions with some 700 T-54/55s to the Syrian front. And although the Iraqi Army waged a static, defensive battle for most of the 1980–88 Gulf war, its first move was an armoured thrust into southern Iran. Its armour, moreover, spearheaded numerous counter-attacks in later stages, and formed a principal element in the 'fire-boxes' that were formed along the front.

The importance of this role was naturally reflected in the activity of the defence industry. Indeed, at first sight it might seem that Iraqi productive projects for vehicles were modest in comparison, though on closer inspection it appears that Iraq undertook extensive work on selected areas of its armoured combat vehicles, putting it ahead of Egypt. Its most ambitious project was the local production of the Soviet T-72, though substantial modifications were also made to T-55 tanks and to BMP-1 IFVs.

• Asad Babil (T-72): the *Lion of Babylon* is a locally-produced copy of the Soviet T-72M1 MBT, of which the first vehicles were completed in spring 1989. Iraqi workshops produced the 125-mm gun, breech and ammunition, as well as some of the tank's optics and electronics.[49] The rest of the tank was imported from the USSR, though it is not clear if sub-systems were sent for assembly in Iraq. However, the ultimate aim of the *Asad Babil* programme was to produce the complete tank locally. As delays and disagreements continued to plague the American-Egyptian M-1 tank deal in 1989, Iraqi-Egyptian co-operation in producing the T-72 was suggested as a possibility.[50] By 1990, the Iraqi Army was operating some 1,000 Soviet-supplied T-72s.

• T-55 *125-mm version*: an offshoot of the T-72 programme was the upgrading of the T-55 of which the Iraqi Army possessed some

2,500 by 1990. Explaining this programme, the head of the Scientific Development Committee, General Ammar, stated that retrofits were seen as a cost-effective method for prolonging the usefulness of weapon systems.[51] Essentially the upgrade involved replacing the D-10T 100-mm gun of the T-55 with the 2A46 D-81 125-mm smooth-bore of the T-72. In order to do so, the turret roof of the T-55 was raised and an ejection door for cartridge cases installed in the turret rear.[52] As the D-81 is automatically-loaded, its use allowed the reducing of the crew by one man to three, though it is not known how the changes affected the number of two-piece rounds carried by the modified tank. An Iraqi-made fire-control system incorporating that of the T-72 is used, along with new passive night sights. Other alterations include the addition of seven forward-firing smoke dischargers on either side of the turret.

• T-55 *upgrade*: a second version of the T-55 upgrade retained the original 100-mm gun. The main improvement was the addition of a locally-made, appliqué composite armour package that protected the turret front, sides and rear and the hull front and sponsons. The same armour consisting of sandwiched steel plates and rubber, covered the front half of the side skirts. Four forward-firing smoke dischargers were added on either side of the turret.[53]

• *Other modifications*: Both the T-55 and the T-72 were provided with a potential anti-mine capability. On the former this took the shape of attachable mine-rollers, while the latter employed a mine plough.[54]

• BMP-1 *IFV*: the Soviet-designed BMP-1 infantry fighting vehicle was the second main armoured combat vehicle to be produced in Iraq under licence. How much was actually produced locally rather than assembled or imported is unclear, but Iraqi manufacture did include at least parts of the hull and armour. Thus the Iraqi version had added rubber and steel side skirts and a new bow plate, which at 35–40mm was thicker than the original.[55] An appliqué armour package was also developed for the BMP-1, providing protection over the frontal arc against 12.7-mm and 14.5-mm rounds at 200m, for a total weight increase of 1,250kg and a slight reduction in the power-to-weight ratio.[56] Further reflecting Iraqi combat experience, the original 73-mm gun may have been replaced in the new version with a 30-mm automatic cannon. Some vehicles also had their roofs raised by 30–40cm in order to operate in the ambulance or the command-post role as well.

Air-Launched Munitions

An instance of the rapid development of the Iraqi defence industry is the production of a variety of guided and free-fall air-to-ground weapons.

- *Guided missiles*: the one guided missile programme in which Iraq was involved at the end of the 1980s was an adaptation of the Soviet AS-14 (*Kedge*) laser-guided missile (designated X-29L in the USSR) for use with the French-made *Mirage* F-1EQ-5 aircraft. The primary modification was to enable the missile to be used in conjunction with the Thomson-CSF *Atlas* laser designator, mounted under the centre line, although the launcher rack and interface also needed alteration.[57] The development, including software changes, reportedly took two months, and permitted an increase in *Kedge*'s range from 6–8km to 10–12km, possibly because of the greater sophistication of the *Atlas* pod.[58] However, Iraq moved after the end of the first Gulf war to obtain a licence for the local production of the French AS 30 *Laser* laser-guided ASM, of which it previously had ordered 586. The project would commence with the assembly of imported parts and then shift to complete local manufacture.[59] (On the *Ababil* ASM, see earlier section on tactical ballistic missiles.)

- *Guided bombs*: most Iraqi effort until 1990 was devoted to developing guided bombs. In production was the *Saqr*, an infra-red guided bomb with a range of 8km. It autonomously seeks any heat-emitting target and may be released at a variable altitude (though not at steeper than 25° diving angles).[60] Under development, meanwhile, was a family of laser- and tv-guided bombs based on a common airframe. This contained four forward canards and a front-mounted seeker (a Japanese camera in the tv-guided version), attached to an ordinary MK84-type 1,000lb bomb, giving an overall length of 5m.[61]

- *Free-fall bombs*: by 1990, Iraq was manufacturing a number of ordinary, iron free-fall bombs of a variety of weights, the largest of which was 9,000kg. Two additional types of bomb were fuel-air explosives and cluster units. Little is known of the former, but the latter included the *Eagle* system (which resembles Chile's Cardoen cluster bomb, sold to Iraq), which contained a variable but undefined payload. The payload probably consists of the anti-tank/anti-personnel bomblets and anti-tank mines developed for the *Ababil* artillery rocket, although none are known to have been employed in 1991. Iraq also produced at least two versions of ordinary air bombs filled with chemical agents.

Other Products

Reflecting the varied needs of the Iraqi armed forces and the growing technological sophistication of the local defence industry, the latter embarked on a number of additional weapons developments.

• Faw-1 *ATM*: the latest of Iraq's local design and development efforts was the *Faw*-1 anti-tactical ballistic missile. It is unclear when development of this weapon started, or if it commenced before the first Iranian use of short- or medium-range ballistic missiles against Iraq, but the first successful test of *Faw* was in November 1988.[62] Iraq also suggested that protection against Israeli ballistic missiles was another aim of this programme.

From its appearance, the *Faw*-1 was a new missile rather than a modified SAM, which implies a higher level of local design input. However, it was to be used in conjunction with an existing early warning radar to detect launches and a tracking radar to determine target data, with control from an intercept ground complex. Had it been a completely new design – as seems likely from its pictures – then the *Faw*-1 missile (if not its ground-based radar system) was comparable to the *Hetz* under joint development by Israel (and the USA). Some *Faw*-1 missiles were already in deployment against short- and medium-range attack, according to Iraqi officials, and a second version was under development for use against longer-range ballistic weapons. However, it may well be that this was an instance of deliberately exaggerated claims designed for foreign ears.

• *Naval mines*: As a result of its extensive confrontation with a dominant Iranian navy, during which it tried especially to block operations at Iranian oil loading and production terminals and rigs, Iraq developed a variety of anti-shipping mines. Most prominent among these was the *Sumer* family, the result of four years' development since 1985. *Sumer* reportedly came in air-, surface-, and submarine-launched versions, and used a modular charge system.[63] The latter was connected to a computer-controlled influence fuze, allowing it to be either moored or ground-laid. According to some reports, the Iraqi *Sumer* may have been based on the Chilean Cardoen *Stonefish* 'intelligent' mine.[64] Other weapons were the *Sajeel* and *Muthanna*-45 shallow-water defence mines, and the *al-Qa'qa'* oil-rig demolition mine.

The Aerospace Industry

At the end of the 1980s, this remained one of the least developed branches of the Iraqi defence industry, whether in terms of the number and nature of its products or in comparison with the relative importance of the industry in other Third World countries (and especially Egypt). But it is also in this sector that Iraqi plans for the future were the most ambitious.

The first stepping stone in the expansion of the aerospace industry was to be the licence-production of an advanced jet trainer, to be followed later by a fighter aircraft. In each case the plan was to seek assembly, then progressing gradually to partial and, finally, complete local production. According to General Sa'di, production of the trainer would rise to 100 per cent of the fuselage, but less than 100 per cent of the engine, unless sufficient investment was made.[65] The process was expected to produce the first locally-assembled aircraft within two years of programme start, with the fighter following. As part of this effort, Iraqi officials also sought acquisition of depot-level maintenance for their aircraft.[66] Iraq had thus chosen a similar path to Egypt and other Third World arms producers towards acquiring a modern, comprehensive aerospace industry, through the gradual acquisition of technology and training. The first phase of this programme – known as the Fao Project – was to have started in 1990, probably based on the French *Alpha Jet*.[67]

Iraqi aerospace efforts up to 1990 focused on modifying imported aircraft already in service in order to meet certain operational requirements, and to manufacture remotely-piloted vehicles (RPVs) and drones. With the exception of the modification of the *Mirage* F-1 to fire Soviet AS-14 *Kedge* laser-guided missiles, mentioned earlier, these changes and developments are reviewed next.

- *Tanker conversions*: an early instance of modification was the conversion of Soviet-built Antonov An-12 transport aircraft to the tanker role, for in-flight refuelling of combat aircraft. Simultaneously, MiG-23 fighters (and possibly other types such as the *Exocet* anti-ship missile-firing *Etendards*) were fitted with refuelling probes in order to extend their range. In spring 1989, Iraq displayed a new tanker that it said had been in service for four years, based on a conversion of the Soviet Ilyushin Il-76 (*Candid*) transport.[68] How much this relied on local development and how much commonality it had with Soviet designs was unclear, since the USSR later revealed its own tanker based on the Il-76, the Il-78.

- *Early warning and airborne command*: another striking Iraqi conversion was also based on the Il-76. This was the *Baghdad*-1 airborne early warning (AEW) aircraft, first shown in May 1989, but reportedly in use against Iran in the final months of the war. The conversion was based on the incorporation of British and French electronics systems, such as the Collins IFF transceiver and the Thomson-CSF *Tiger* surveillance radar.[69] The latter, also produced locally in a land-based towed trailer version, was mounted under a GRP radome that replaced the belly cargo door of the Il-76, giving it less than the 360° coverage of American and Soviet types but 'much more than 180°,' according to Iraqi officials.[70] The same sources stated that power was provided by a locally-designed generator, and that information was transmitted by a real-time data link or voice. The radar system, with a crew of four, could reportedly detect, track, and identify an unspecified number of targets at ranges of up to 350km. In order to adjust handling, two strakes were added under the rear of the aircraft.

At the time the *Baghdad*-1 was displayed, Iraq was also reported to be working on an airborne command version, the *Baghdad*-2, to control fighter aircraft in combat. However, the whole series was subsequently cancelled, in favour of a new version of AEW/AWACS aircraft. This was revealed to be the *Adnan*-1, which was again the Il-76 but fitted with a rotodome (rotating radar) instead of the belly-mounted *Tiger* of the earlier series, in order to give better coverage.[71] Despite close resemblance to the Soviet *Mainstay* aircraft, Iraqi officials confirmed that the aircraft and system were 'integrated' within the country – a considerable technological challenge.[72] Radar detection was reported to be 'several hundred kilometres' (radar dish size is 9m in diameter, mounted 4m above the fuselage). Two *Adnan*-1 aircraft flew to refuge in Iran during the second Gulf war.

• *RPVs and drones*: Iraq has exhibited four types of RPV and a target drone. Of the former, two appeared to be in serial production by 1990 – *al-Yamamah* and *Marakeb*-100 – while the status of the remaining two – *Sahreb*-I and *Sahreb*-II – was unclear. *Al-Yamamah* was a locally-produced, mini-RPV with a 50 km radius and a maximum speed of 170km/h, of which development started in 1986 and prototype production in 1987.[73] *Marakeb*-100, conversely, was a copy of the *Meteor Mirach*-100, with local manufacture of the fibreglass body and use of imported engines and electronics. Finally, a target drone (designed as a mockup of the Israeli *Kfir* C-2) has been displayed.

• *Electronics*: The electronics factory established by Thomson-CSF near Mosul in 1980 has provided a variety of components for other military products. Among these may be some of the IR and laser sensors used on air-launched guided bombs, and cameras for RPVs. Software and interfaces for air-launched munitions, such as the AS-14/Mirage F-1 match, are other possible items, and certain components may be licence-produced or hybrid sub-systems designed for use on the *Baghdad*-1/2 and Adnan-1 AEW and AC aircraft, for example. Electronics work is probably undertaken using a combination of imported technologies and hardware and local design and adaptation. An example is the use of US electronics and computer software supplied by Hewlett-Packard in 1985–1986, in the Saad-16 complex.

The major system produced by the Iraqi electronics industry is the *Tiger* G low-altitude air defence radar, built under licence from Thomson-CSF. Known as the SDA-G, it is trailer mounted, though a modified version is produced for operation on the *Baghdad*-1 AEW aircraft.

• *Nuclear and chemical weapon programmes*: Several states in the Middle East are known to possess nuclear or chemical weapons, or at least to have the capability of producing them at short notice. Israel has both types in its arsenal, while Egypt is known to have used chemical weapons during its intervention in the Yemen in the 1960s. Egypt, Syria and Libya are moreover reported by US and Israeli sources to be currently engaged in the manufacture of chemical weapons, using primary materials purchased from a variety of sources as far apart as (formerly West) Germany and India.

However, Iraq is the only Arab country known since the mid-1960s both to have produced and employed chemical weapons *and* to have been developing a nuclear weapons programme at the time of the 1990 Gulf crisis. Indeed, official Iraqi statements issued after the end of the war of 1991 admitted the existence of extensive stockpiles of chemical munitions and of previously undetected nuclear facilities and quantities of fissile material. Western sources estimated that Iraq had invested $10bn in the nuclear programme alone.[74]

Before 1990, Western sources had stated that Iraq had two main complexes involved in chemical weapons production: a large one at Samarra (26 square km in area), 100 km northwest of Baghdad, and a smaller one at Faluja, some 65 km to the west. A research centre (also rumoured to be involved in germ warfare research) was based at Salman Pak, 56 km southeast of the capital. The same sources claimed that these factories produced older agents – mustard gas and lewisite – and new nerve gases – tabun and sarin – with manufacture starting in 1984.[75] The Faluja plant was supposed to be producing precursor agents for Samarra, and annual production was estimated at 780 tons of the older blister agents and 50 tons each of sarin and tabun.[76] The latter plant came officially under the State Establishment for Pesticide Production.

On 20 April 1991, following the second Gulf war, the Iraqi Foreign Ministry submitted a letter to the UN Secretary General containing a list of chemical weapons. These included 6,620 war-heads for 122-mm artillery rockets and 2,500 warheads for *Saqr*-30 rockets, and 600 DB-2 aircraft bombs, all containing sarin and stored at the Muthanna General Plant (70 km west of Baghdad).[77] This plant also comprised 25 laboratories, plants for producing sarin, tabun, mustard gas and intermediate chemicals, munitions-filling lines, and 955 tons of the three agents mentioned. An additional 105 155-mm artillery shells packed with mustard gas were at the Chemical Training Camp near Faluja, while 1,040 aircraft bombs filled with mustard gas and 336 binary system bombs filled with sarin were stored at six airbases around the country. Thirty chemical warheads for *Husein* ballistic missiles were based at Dujayl. There was no indication that any other artillery systems (mortars or guns/ howitzers below 155-mm calibre) could fire chemical munitions.

In July 1991, the Iraqi authorities also made declarations indicating the existence of a nuclear weapons programme before the Gulf crisis of 1990 (though they continued to deny any biological capability). This revealed the presence of a number of facilities that had not hitherto been detected or reported, and of 90lb of uranium, in addition to a similar amount known to be in Iraqi possession.[78] Earlier, the press had listed 24 facilities involved in aspects of nuclear R&D or related activities, and confirmed that Iraq had 27 kilograms of enriched uranium.[79] Not all facilities were destroyed during coalition air attacks during the second Gulf war, though UN investigators were able to inspect most sites and tally the fissile material. It remained unclear how close Iraq actually was to producing any type of nuclear device, but most significant was its ability to move independently, using Western inputs that it then adapted and combined, to start developing its own centrifuge process for the enrichment of uranium.[80] Iraq was successful in obtaining calutrons, enabling inefficient uranium enrichment. Iraqi claims of indigenously producing krytron nuclear triggers were probably untrue, though steps were taken in this direction.

Organizing Defence Production: The Western Role

Serious doubts might be cast on the extent to which local scientific input and manufacturing know-how contributed to the development and production in Iraq of the weapons, munitions, and support equipment described in previous sections. And questions might also be posed concerning the real volume of output and its quality and regarding the actual status of systems claimed to be in production but probably still in the development stage. What cannot be denied, though, is that Iraqi ability to undertake military industrialisation, to the extent that it has, reflects a high degree of political will and determination, coupled with an equally notable organisational capability: that is, the capacity to identify needs, set targets, and marshal resources.

A particular case of Iraqi success in the defence sector has been the involvement of a large number of countries and companies in the local industrial effort during the 1980s. In addition to Egypt, North Korea, and Yugoslavia, the role of Western experts and companies was critical in Iraqi development and production programmes. As significant was the fact that the relevant Iraqi government agencies had both the funds and the authority to establish an international financing and procurement network, run either from Baghdad or through Iraqi-owned companies based abroad. The collapse of Banque de Crédit et Commerce Internationale in 1991 revealed the manner in which Iraq, among other countries, had financed part of its procurements. Whether for this reason or a desire to make profits, it cannot be a foregone conclusion that no Western company or other foreign agency will resume collaboration with the Iraqi military industry, should it resume activity in the wake of the war of 1991.

Western input – mainly private – into Iraqi military industrialisation has been in both the conventional and the non-conventional spheres. This has often involved the sale of machine tools, raw materials, and components, as well as technical assistance and technology: some of the sales were possibly designated officially for the civilian sector but were diverted to the Iraqi arms industry, unkown to the original suppliers. Alarmed at the scope and scale of this illegal trade, the USA drew up an international blacklist in Spring 1991 of companies involved in supplying Iraq.[81]

According to reports, a prominent example of the role of Western companies is that of Matrix Churchill in England. It apparently was on order to supply Iraqi industry with £19m worth of lathes, including 200 computer numerically controlled lathes, that could be used in the manufacture of shells, and for another shipment for the production of fuses.[82] The lathes were to be constructed in Iraq from kits, and installed in a shell-making factory set up by the Chilean company Cardoen. Matrix Churchill was also reported as receiving a loan of £26m – from an American branch of the Italian Banca Nazional del Lavoro, which had given Iraq vast but unauthorised loans variously estimated to total from $1.7bn to 3bn – to build a die-forging plant on a turnkey basis, officially for the civilian automotive industry run by the Nasr State Enterprise for Mechanical Industries.[83]

Western media also reported that such English companies as Colchester Lathes and BSA supplied Iraq in the late 1980s with lathes and jigs, some of which went to the weapons and explosives testing site at Qa'qa' (south of Baghdad) and to a shell-making plant at Badr.[84] The West German firm SPS-HPG won a contract to design and manage a five-line forging plant for Nasr, though the company stated that this was for the civilian motor industry.[85] These were only some of a larger number of companies involved in similar trade.

A further specific example of the foreign role in Iraqi military industrialisation, and of the way in which Iraq managed to organise partial inputs from different companies to produce complete manufacturing projects, is that of Cardoen. This was claimed to have commenced supply of cluster bomb technology to Iraq in co-ordination with the American company International Signal and Control (ISC), starting in 1983, although Cardoen denied the connection.[86] Western sources assert, however, that Cardoen did provide Iraq with a turnkey facility for the manufacture of cluster bombs, for which ISC supplied charts, bomb and fuse designs, and other know-how, while Matrix Churchill supplied the lathes.[87] The West German firm MBB was meanwhile reported to be the source of FAE (fuel-air explosive) bomb technology: this was the product of an R&D effort in Egypt by MBB, which was passed on to Iraq by Egypt in 1988 when MBB suspended its involvement.[88]

Some of the same companies were also involved in supplying the Iraqi development and production effort in non-conventional weapons. Matrix Churchill and BIMEC Industries were both reported to have supplied Iraq with the jig-boring and metal-plating machines used to extend the outer casing of its *Scud* missiles and to coat missile components (for heat- and corrosion-resistance), as well as other machines to modify internal components such as the rocket motor.[89] Additionally, the Belgian-based Space Research Corporation is supposed, in addition to its involvement in the super-gun affair, to have sought filament winding machines, which can be used in the manufacture of missile nose cones, while Austrian and West German firms designed and built a missile-testing wind tunnel at the Saad-16 complex near Mosul.[90] Prominent in the last effort were Gildmeister Projecta and MBB, which either acted as contractors, provided R&D, or were the source of former personnel.[91] Iraq also tried to obtain assistance in the production of stabiliser fins, testing equipment and materials, as well as hot isostatic presses for the processing of carbon-carbon, used in making rocket nose cones and engine nozzles.[92]

Finally, Western input was of great importance in the development of the Iraqi chemical and nuclear programmes. The prime suppliers in the former case were former West German companies, some of which commenced the export of precursor chemicals and machinery for the production of gases – 17.6 tons daily of tabun at the Faluja plant – and chemical shells in 1986.[93] The firm concerned here was Water Engineering Trading (WET), which also provided the Salman Pak and the Samarra plant with equipment of potential use in the development of biological weapons; according to German sources, other companies sold 200mg of T-2 and HT-2 mycotoxins to Iraq, aircraft drop tanks for chemical weapons, and built six chemical factories (one of which was later confirmed as a weapons plant).[94]

In the nuclear sphere, Western companies provided machines that could be used in the production of fissile materials. These included high-speed rotors, computerised testing machines (permeators), precision lathes and magnets for the manufacture of centrifuges, which could be used for the enrichment of uranium, and non-corrosive steel designed to withstand uranium hexafluoride inside the centrifuges.[95] Other items included 'skull' furnaces capable of melting plutonium or uranium for the cores of nuclear bombs, or titanium for missile nose cones – export of which was prevented by the American and the British government – and 'krytrons', special electronic triggers of the type used to detonate nuclear devices.[96] British, American, and West German firms were especially active in this trade.

Assessment

There is no denying that the Iraqi defence industry made impressive achievements in the space of a few years. This rapid growth was even more impressive partly because it occurred during the war with Iran, and partly because the military production sector was youthful and could not draw on a broad and experienced civilian industry. By way of comparison, the modern military industrial tradition in Egypt extends back for at least four decades. Thus qualifications may be made about the extent of Iraqi dependence on foreign technology and assistance, or about the real proportion of originality in local design and manufacture, but the local industry appeared to be as developed as that in comparable Arab and Third World countries.

That said, it is relevant to assess the specific mix of local and foreign inputs and to evaluate the potential viability of the Iraqi defence industry on the eve of the Gulf crisis, in order to highlight the issues involved in Arab and Third World military industrialisation. At that point in time, Iraqi military products could be divided broadly into two main types. One was based on copies or on imported designs and licensed production of foreign weapons and systems. The second was hybrid designs resulting from Iraqi modifications or joint efforts between Iraqi and foreign designers, with varying degrees of actual Iraqi manufacture. Indeed, in some cases it appears that the proportion of Iraqi input at all levels of system design, development, and production was high enough to place certain systems in a third category of basically indigenous military products, although information is still insufficient to confirm this.

It is relatively easy to specify Iraqi products that are copies or based almost wholly on imported designs. Most infantry weapons and most categories of conventional, unguided munitions fall into this category, as do such weapons as 60/82/120/160-mm mortars and the 122-mm D-30. The same can be said of the T-72M tank and the BMP-1 IFV, although the latter contains some Iraqi modifications. Furthermore, even though they may be produced for use in new combinations, many sub-systems and components, such as electronics or 155-mm gun and 125-mm tank gun barrels, were produced under foreign licence. These licences are extremely unlikely to be renewed following the war of 1991, but Iraq may yet succeed in overcoming at least some obstacles once the central government rebuilds its foreign relations and the former Soviet republics and other countries move to regain some influence in the Middle East.

It is also relatively easy to specify those Iraqi products that are not simply near-exact replicas of imported weapon systems, regardless of who was actually most responsible for the new designs: the 155mm conversion of the 130mm M-46 gun, *Fao* and *Majnoon* self-propelled guns and 120-mm mortar carriers, upgunned T-55, *Ababil* MRLS, *Husein/Abbas/Waleed* family of tactical and medium-range ballistic missiles, *Faw* ATBM, laser- and TV-guided aerial bombs, *Sumer* naval mines, and Il-76 tanker and AEW versions. In each case, the system concerned is, as an integral whole, new or at least substantially different from existing equivalents, and has thus involved a significant degree of additional design work, and possibly also of modified manufacturing techniques. The same also applies to certain components, such as the cluster warheads for the 122-mm D-30 gun and *Ababil* rocket or the 210-mm gun barrel for the *Fao*, which are not in service elsewhere.

Within this broad category, some systems are obvious developments, to lesser or greater degrees, of existing weapons. In certain cases, this involves combining several sub-systems and components of foreign design or make to produce a new system. For example: the new version of the M-46 gun combines a French-designed 155-mm barrel with the original Soviet trail; the earlier 120-mm mortar carrier mounts a standard barrel inside an MT-LB vehicle, both of Soviet origin; the upgunned T-55 mounts a 125-mm gun from the T-72; and the Il-76 (*Baghdad*-1) AEW aircraft uses French electronics. Of course, achieving these hybrids is no mean feat technically, and has moreover necessitated additional adaptations, such as altering the T-55 turret to accommodate the 2A46 D-81 125-mm gun.

In other cases, the end-product is more than a combination of disparate components, and has involved more basic engineering redesign than simply getting sub-systems of different origins (such as Soviet laser-guided missiles and French designators) to work together; examples are: the *Majnoon* 155-mm self-propelled gun, which though similar to the South African G-6 is based on a new articulated vehicle; the *Husein* and *Abbas* tactical ballistic missiles, based on the Soviet *Scud* but with markedly different payloads and ranges; the An-12 and Il-76 inflight refuelling tankers (and MiG-23 refuelling conversions); and the four-barrel, 120-mm mortar carrier involving extended tubes and an externally-mounted hydraulic lift system. (The example of the *Majnoon* articulated vehicle shows that the design may result from foreign input – in this case Spanish – but that the end result is new.) An intermediate case is the new TV-guided glide bombs using Japanese cameras, which nonetheless involve new aerodynamics as well as interfacing with the aircraft. Iraqi industry, moreover, made minor but original contributions to other weapons (with or without assistance), such as the appliqué armour on the T-55 and BMP-1.

At the other end of the scale, other systems may owe much to foreign technological assistance (whether in design, development, or production), but nonetheless are completely new weapons. The more modest cases are the *Fao* 210-mm self-propelled gun, which employs the largest artillery calibre now in service worldwide, and the new family of laser-guided glide bombs. The preceding are somewhat familiar in their general design and origins. But the *Faw*-1 ATBM, *Waleed* MRBM (both still under development or in early production stages), *Ababil* 400-mm MRLS, and *Sumer* mines are all completely new. This applies to their design, aerodynamics and engineering, guidance, propulsion, and combat payload.

In assessing this record, it is virtually impossible to ascertain exactly how much of the input to design work on these new or modified systems is local, in the absence of official information from Iraq or from its former foreign partners. The conventional view of the Iraqi defence industry up to 1990 was that it lacked military technology and trained personnel.[97] Iraq, moreover, had a relatively weak civilian industrial sector, whether in terms of sophistication, output, and local technological contribution (in design and manufacturing techniques), or in comparison with both population size and economic resources, or to, say, Egyptian industry. Besides, the role of foreign assistance (official or private) in producing some weapon systems is an established fact: Yugoslavia in the 122-mm D-30, France and other European countries in the *Fao* and *Majnoon* SP guns, and the former West Germany and Egypt in the *Husein* SRBM. And given the role of such companies as France's Thomson-CSF in establishing the Iraqi electronics industry, additional foreign co-operation may have occurred in developing hybrid or new systems based on components of outside origin.

Conversely, there were an estimated 100,000 persons employed in the Iraqi defence industry by 1990, which points equally to massive local involvement and to the availability of personnel of at least minimal technical capabilities. But sheer scale is no guarantee of effectiveness, or that technology was indeed being transferred by on-the-job training. The Iraqi record in acquiring know-how through the importing of designs and skilled personnel has not been impressive in the civilian sector, either in industry or in construction. On the other hand, the security sensitivity of the military production sector and the urgency of the need for reasonably efficient industrial performance (not to mention dependable end-products), due to the war, may have stimulated better results.

In fact, urgency works both ways. It may have prompted less emphasis on achieving greater autonomy or self-reliance and more on quick material results for use in the battlefield against Iran, thus in its turn prompting indiscriminate recourse to foreign experts and expertise. It may also have led to low cost-effectiveness, in the rush to obtain combat equipment. Whether or not the local industry acquired enough experience and know-how in the process would have become more evident in the period after the war with Iran, both as information became more available and as the Iraqis attempted to launch new programmes and attract export customers. Their ability to keep pace with technological developments, improve products, and achieve competitiveness internationally depended on just how much their personnel actually absorbed during the experience and contact of the preceding few years. Obviously, Iraqi capability might not be put to the test publicly.

On balance, however, it appears that urgency stimulated real development in the defence industry in the 1980s. One of the most striking features of Iraqi military products is the degree to which they reflected combat experience and were accordingly tailored to local needs. This means, at the least, that Iraqi personnel were able to evaluate performance and assess its implications for the equipment, and then to issue precise technical specifications for alterations. The Iraqi Army and defence industry, partly through the medium of the Scientific and Development Committee, therefore achieved the type of interaction so necessary to successful R&D, a channel lacking in the Egyptian case in the 1950s and 1960s and still flawed since. Thus the fact that many new Iraqi systems are little more than modifications of older weapons or hybrids using established components is in reality a positive indication. In many Third World countries, and even in such states as Israel and South Africa with their developed industries, it is often the ability to adapt weapons systems to local needs that is most rewarding operationally, technically, and financially. The difference in each case is how much of this adaptive ability is indigenous.

Whatever the final judgement on the balance between Iraqi and foreign inputs to the defence industry, it does not necessarily affect the separate question of the commercial feasibility of local arms production. It may have been necessary to produce *matériel* even at inflated cost during the war with Iran, but this is something that Iraq could not afford in peacetime, and certainly not after the second Gulf war. The financial and economic consequences of the latter conflict will weigh heavily on Iraq throughout the 1990s, at the very least.

However, this may also suggest that the types of product that the Iraqi military industry can produce – combat consumables in particular – despite sanctions and indebtedness, will depend less on foreign technological input than will other products such as aircraft.

Prospects for the Iraqi Defence Industry (Pre-1990)

Before the Gulf crisis of 1990, Iraqi plans for the development of the local defence industry relied on three parallel courses of action. These were: to develop indigenous technological and productive capabilities, co-operate with other countries in order to establish new programmes and acquire technology, and expand associated civilian industries.

The first option related to the establishment of a new R&D bureaucracy and the expansion of existing ones, especially those engaged in highly sensitive areas of military production. Besides the nuclear and the chemical branch, this may have included space technology, ballistic and anti-ballistic missiles, and specialised conventional items, such as armour for AFVs, precision-guided munitions, and cluster warheads. Of course, much of this effort still involved foreign participation, by Third World countries such as Brazil or by Western private companies, but the emphasis appeared to be on actual transfer of technology to Iraqi research teams.

Co-operation with other countries, secondly, offered several advantages. The most immediate was to introduce Iraqi industry to branches of activity that it was weak in, principally (but not exclusively) aerospace. Thus discussions with France, started in February 1989, revolved around possible Iraqi assembly, then production of the *Alpha Jet*, with an eye to eventual assembly and manufacture of the *Mirage*-2000. Another possibility was that Egypt would help Iraq to establish assembly and production lines for the *Alpha Jet* and the *Gazelle* helicopter. A third option under discussion was obtaining a Brazilian licence for the local production of the *Tucano* trainer, again with the possibility of benefitting from Egyptian assistance. The USSR was another potential partner, making a strong bid for military industrial co-operation in the same period. Among the many added attractions of this option were: Iraqi familiarity with Soviet weapons, the dependability and ruggedness of the latter, cheapness compared with Western products, and greater willingness to envisage technology transfer.

It is evident that in these cases the importance of co-operation with foreign partners lay largely in the developing Iraqi ability to produce main weapon systems. These are extremely varied and sophisticated technologically, involving hundreds and even thousands of subsystems. This does not imply that Iraq was seeking complete production; the strategy was clearly 'to produce parts or assemblies that will best contribute to our independence and freedom of action . . .

so we have made an attempt to concentrate our resources in certain areas'.[98] This form of co-operation was therefore seen as a means for acquiring those areas of technology. The same applies to other Iraqi programmes. In some cases foreign input might have been directed to developing entirely new systems, while in others it would have taken the form of licensing Iraqi production of medium weapons and sub-systems. In particular, Iraq might have wished to turn to licensed production of weapons it did not already produce – such as ATGMs, AAMs, and man-portable SAMs – and specialised areas, such as communications, computers, and optronics. Finally, another advantage of foreign co-operation in weapon systems would have been the possibility of boosting Iraqi export potential, whether of whole systems or of re-exported components produced under licence.

The third course of action was the constant effort to expand and develop the civilian industrial sector, especially in such defence-related areas as steel production and electronics. On the one hand, this would have the effect of feeding the defence industry. On the other hand, expanding the civilian sector would supposedly enable it to benefit from the spin-off effect of military industrialisation. This was especially so in a situation like the Iraqi one where, owing to the primacy of the war effort in 1980–88 and the continued major impact of security considerations or of Iraqi regional ambitions after the ceasefire, the defence industry was the recipient of lavish funding and large numbers of trained personnel. In any case, as important an incentive for this course of action was the need to reduce dependence on basic primary materials and intermediate components necessary for defence production, by encouraging their manufacture locally.

Developing associated civilian industries was an important element in achieving economics of scale, moreover, as military consumption of any item (steel products, for instance) could probably not alone be sufficient to make local production feasible. This does not mean that economies of scale would automatically result from the expansion of the civilian sector; the civilian sector might not itself be able to generate enough projects or consumption. Nor, in the absence of greater private sector involvement or of markedly improved performance (compared with the past record), were new manufacturing projects likely to increase feasibility. In any event, Iraq soon embarked on establishing a large number of new industrial enterprises after August 1988, some in partnership with local private or foreign capital. An example is the Iraqi-Egyptian deal signed in March 1989, worth $240m, for the construction in Iraq of two steel and iron manufacturing plants intended to supply products for the manufacture of aircraft, ships, vehicles, and machine tools.[99] This was part of a broader plan to raise local steel output to 5m tons annually.[100]

Epilogue

The aftermath of the Iraqi invasion of Kuwait in August 1990 has cast a wholly new light on military industrialisation in Iraq, as indeed it has on similar Arab efforts. The Iraqi arms industry was extensively damaged during the Gulf war and will suffer a shortage of funds as the country pays punitive reparations in years ahead. Sanctions imposed by the UN and others may also constrict the flow of components, tools, and military technology to the sector, in addition to the limitations that might be set on the size of the Iraqi armed forces and on the types of the weapon they are allowed to field. The UN Security Council resolution passed in April 1991 banned the supply of anything relating to the design, development, manufacture, and use of weapons, among other restrictions on NBC and ballistic missile programs.[101] For its part, the US American Treasury meanwhile compiled its own blacklist of companies and persons said to be assisting 'Iraq's subterranean network in the world of arms, trading and clandestine financial operations'.[102] The Congress later demanded a review by all relevant government departments of the transfer of technology to Iraq over the previous decade.[103]

Nonetheless, a number of compelling factors may lead Iraq to renew its efforts at military industrialisation, despite comments by the Minister for Industry and Military Industrialisation, Amer Mahmoud Saadi, that there can be no thought of arms production with so much damage to the economy and that the sector's capacity would be turned to civilian use.[104] Although the restrictions imposed on Iraq are both severe and sweeping, they may not entirely impede reconstruction of at least part of the local arms industry. Indeed, the embargo may spur an even greater emphasis on indigenous production in order to raise autonomy, however slightly. Certain sectors might prove easier to refurbish and resupply: those that involve low-tech manufacturing processes or products (ammunition and certain spare parts, for example), or that overlap with civilian industry (such as electronics).

Furthermore, a number of major arms producers have a vested interest in maintaining at least part of the Iraqi market, whoever is in power; private companies in the West or sympathetic third countries might also assist in the transfer of military technology to Iraq. The former Soviet republics in particular are likely to show themselves sensitive to the defence needs of such Arab states as Iraq and Syria, partly for political reasons, while a range of countries from China and North Korea to Brazil, South Africa, and the East Europeans need export markets. Should domestic peace and stability be restored in Iraq, along with reforms that permit a lifting of UN economic sanctions and resumption of trade with the outside world, then maintaining sizeable armed forces and related productive facilities may remain both attractive and feasible as a practical option.

6

Saudi Arabia: One Foot Forward, One Foot Back

Saudi Arabia is the third Arab country, after Egypt and Iraq, with the potential to become a principal producer and ultimate exporter of arms regionally. In terms especially of the availability of surplus funds, which is reflected in the nature of its military industrialisation plans discussed below, Saudi Arabia is well placed for such a transformation – at least on paper. Yet after several false starts the kingdom is moving only slowly at the beginning of the 1990s towards its targets.

The Saudi case is particularly striking because of the contrast between the abundance of funds to finance military industrialisation and the relative scarcity of most other requirements: non-petroleum natural resources, civilian manufacturing industry, industrial 'culture', and a large industrial workforce of varying skills and levels.

Moreover, despite being a major power in a tense and highly volatile region, Saudi Arabia had never been directly threatened until the Iraqi invasion of Kuwait in August 1990, and so lacked the direct incentive felt by Egypt or Iraq to establish a large defence industry. Indeed, this is strong evidence of the primacy of national security goals as an incentive for setting up local military industries in Arab (and other developing) states.

At the same time, neither did Saudi Arabia have the usual economic and financial incentives to establish a defence industry: it hardly lacked hard currency or exportable products, could afford to set up its civilian sector in any manner it pleased without need of the supposed spin-off effect from the defence sector, and had little problem employing or subsidising its workforce.

Following the Gulf war of 1991, Saudi perceptions may change, with the experience imposing a rethink of national security and defence policy. This may elevate the importance of local defence production in the minds of Saudi leaders, the more so if it promises to allow savings to be made at a time of declining Saudi financial wealth.

Aims of Saudi Military Industrialisation up to 1990

Saudi Arabia was among the Arab countries that first considered practical steps towards establishing a joint defence industry in the first half of the 1970s. It was the other major participant, along with Egypt, in forming the four-member Arab Organisation for Industrialisation in 1975, and therefore shared that body's main aims. As described earlier, the latter were to: (1) achieve a measure of self-sufficiency in conventional hardware (weapons, ammunition, and spare parts), and enable Arab armies to fight longer; (2) promote inter-Arab co-operation and integration, and limit Israeli or other political pressures; (3) reduce the unit cost of military production by extending production runs, and avoid redundancy; (4) provide a basis for advanced industrialisation in the Arab world and train technical manpower; and (5) provide additional income through sale of surpluses to other Arab, Islamic, and Third World countries.

The collapse of the AOI in 1979 evidently led to a Saudi reappraisal of its own aims in military industrialisation. At the very least, the experience demonstrated the risks involved in relying on the establishment – with Saudi financial assistance – of a defence industry based physically in another country. It also ended, for the foreseeable future, any plans for inter-Arab industrial co-operation in the defence sector and altered fundamentally the economics of production. Saudi Arabia now faced the prospect of building up a local defence industry from scratch, a proposal that not only involved major investment but, moreover, would have to depend on an underdeveloped manufacturing base and an inadequate technical manpower pool. The alternative to the latter would be to rely so heavily on expatriate technicians and imported technology as to make the defence industry essentially an imported one.

Indeed, these considerations underlay Saudi insistence in the early 1980s on developing a new plan for military industrialisation. Driving this was the sense of vulnerability to external pressure through an overwhelming reliance on imported labour and hardware, coupled with awareness of economic over-reliance on a single export commodity (oil). In the first instance, this sort of vulnerability demostrably led to insecure military supply. A prime example in the period 1979–83 was the suspension of American arms exports to neighbouring Jordan because of the latter's refusal to join the Egyptian-Israeli Camp David peace process. Saudi Arabia itself has repeatedly faced severe difficulties in obtaining needed items from the USA owing to strong lobbying in the Congress by pro-Israeli members and the American-Israeli Public Affairs Committee (AIPAC). In the second instance, despite heavy investment, Saudi industry accounted for under 5 per cent of the kingdom's GDP by the mid-1980s, while oil and gas and refined products accounted for 50 per cent. Thus dependence on oil exports, and vulnerability to the state of world demand and supply, pose potential problems for Saudi Arabia.

Reflecting these concerns, Saudi Arabia initiated a new programme for indigenous military industrialisation in 1983/84. Among its primary aims were acquiring technology transfer and diversifying the economy.[1] Local development would also permit a reduction in the number and scope of foreign personnel and suppliers involved in sensitive defence sectors. In parallel, the establishment of a Saudi defence industry would enhance the kingdom's political prestige and regional status. However, although the aforementioned technical and economic considerations were both real and of large import, there was a further underlying aim to the whole programme. This was expressed implicitly in August 1984 by the Minister of Defence and Aviation, Prince Sultan bin Abdul-Aziz, who linked current industrial projects with the need to avoid pressure from supplier countries that were themselves 'subject to Israeli pressures' (a reference to the USA).[2]

Saudi Military Industrialisation Strategy

Following its withdrawal from the AOI in mid-1979, Saudi Arabia activated plans for the establishment of local ordnance factories designed to meet its needs for such basic items as infantry weapons and ammunition. The first such project was the construction of a plant by the German Heckler and Koch company to produce its G-3 assault rifle and ammunition under licence. This was based at Kharj, 60km south-east of Riyadh, which was intended to be the main site of the nascent arms industry. A second factory for the production of tank and artillery ammunition was to be established there too, by a consortium consisting of the German companies Thyssen Rheinstahl Technik, Rheinmetall, and Wasag Chemie. This was blocked by the German government, however, which cited its policy of prohibiting the sale of arms and military technology to regions in conflict.[3]

Despite this setback, Saudi Arabia was able to expand its ordnance factories in the mid-1980s for the licensed production of American M-1 rifles and M-16 assault rifles and ammunition for the M-60 squad and 0.5-in heavy machine guns.[4] Even before these entered full operation, though, Defence Minister Prince Sultan was already claiming in August 1984 that the kingdom was self-sufficient in small arms (referring to the H&K G-3).[5] Further expansion plans at the

Kharj complex envisaged the construction of five additional ordnance factories in the second half of the 1980s.

Nonetheless, ordnance formed only a part of Saudi needs, leaving major areas unattended. Moreover, as Prince Sultan declared in 1984, Saudi Arabia wanted to boost production and make new weapon types, as well as parts for electronic systems.[6] Talks were initiated with Brazilian manufacturers, among others, resulting in a Saudi-Brazilian protocol for co-operation in the exchange of technology and arms production.[7] Among the items supposedly under consideration were the Brazilian *Astros* MRL system, *Osorio* MBT, and *Tucano* trainer aircraft, but nothing ever emerged from the agreement.[8]

But given the inadequacy of the existing local industrial base and technical workforce, no expansion into high technology areas – such as electronics, armour, and aerospace – could be envisaged along previous lines. What was needed was a dual process by which, in addition to the immediate objective of actually setting up production plants to meet local military requirements, Saudi Arabia could both acquire technology and establish the industrial base necessary to absorb it and provide for future expansion.

The way chosen out of this seeming impasse was to involve foreign companies massively in Saudi military industrialisation: indeed, to create a local defence industry almost entirely through their efforts. On the face of it this strategy was not unlike the setting up of joint ventures in Egypt between the AOI and Western companies in 1975–78.

The differences were in fact major. The approach of the AOI (both before and after 1979) brought state capital (first Arab and then solely Egyptian) together with private Western capital, in order to establish productive plants actively involved in the assembly and manufacture of military equipment. Hypothetically, the AOI encouraged private Arab capital and offered a degree of freedom from government control, but in practice virtually none of the former was forthcoming and Egyptian state control of the industry remained effectively absolute. Moreover, the investment capital of the joint ventures was relatively modest, and except in limited cases (such as the sub-contracted production of *Mirage*-2000 parts for re-export to France) the government's military transactions did not include a major commercial component on behalf of the defence industry companies.

The new Saudi strategy contrasted sharply with the Egyptian approach. Its main features became evident in March 1985 when Saudi Arabia and the USA finalised the 'Peace Shield' programme, worth $3.7bn, for the supply of a complete air-defence system. A wholly new element in the contract was an offset clause. This stated that American companies bidding for orders would have to re-invest capital – equal to 30–35 per cent of the hardware component value – in industrial projects in the kingdom. As the hardware value (excluding construction and engineering) was roughly $1.5bn, total offsets were supposed to reach $500m.[9] The Saudi Industrial Development Fund (SIDF) was committed to provide a similar amount, on the basis of setting up local joint companies, in which the Saudi stake of equities would be 60 per cent for 5 years and then rise to 70 per cent.[10] In this context, SIDF policy is to offer grants worth up to 50 per cent of the establishment capital of new manufacturing enterprises, with at least 50 per cent coming from domestic or foreign private capital.

The Peace Shield programme eventually led to the approving of four separate projects, worth $600m, presented by the consortium led by Boeing and General Electric.[11] While these were in the process of being set up, Saudi Arabia concluded major new arms deals. In February 1986 came Yamamah-1, in which the kingdom ordered 72 *Tornado* fighter-bombers and other equipment from Britain at a total programme cost of £5bn. This deal contained an offset provision – which was not contractual – for the re-investment of an amount equivalent to 35 per cent of the technical content of the total project cost, or some $850–1,020m.[12] The British government was reluctant to implement this clause, though, which remains unfulfilled. Saudi Arabia nonetheless insisted on a more definite, though still not contractually binding, offset element (worth 25 per cent) during subsequent negotiations over the follow-on Yamamah-2 deal, which was concluded in October 1987.

The offset content in the second Saudi-British deal (worth £10bn) was reported to be worth some £500–600m, but the details were left vague and the British commitment was not binding. Further talks finally led to the signing of an offset memorandum on 23 November 1988, according to which the British government was to encourage private companies to invest a total of some £1bn in Saudi Arabia.[13] This was supposed to equal approximately 35 per cent of the value of the British hardware component (provided largely by British Aerospace) in the Yamamah-2 arms deal (as some of the *Tornado* components and other systems were not of British make).[14] One option suggested for re-investment was to form joint ventures, with the Saudis providing half the capital. In all cases, the offset did not have to be in productive projects nor in defence-related fields; instead it could involve both civilian projects and commercial activity such as the granting of licences, technical training, and expanding export sales for Saudi companies.[15] However, despite the rise in the total offset value agreed, the time span for implementation was extended almost indefinitely.[16]

The progress made by Saudi Arabia towards achieving military industrialisation under these two main offset agreements will be discussed later, but here it is worth recapitulating the main points in its strategy in order to complete the comparison with the Egyptian case. The first and foremost aim of Saudi planners was to build an advanced, technology-based industry funded by offsets. Egyptian industry, in contrast, formed joint ventures and sought production licences but did not ensure sales through offset arrangements. Furthermore, the Saudi industry thus created was to be overwhelmingly privately-owned; funds from the state-run SIDF did not necessarily signify state control over the joint ventures, as the equity stakes were to be offered to local entrepreneurs and eventually converted into high concessionary loans.[17] A further implication is that the general role of the state in directing defence industry activities was to be minimal, unlike the Egyptian situation.

Another difference between the Saudi and the Egyptian case lies in the nature of offset arrangements and how they were made. Major Egyptian arms deals have, in fact, contained little or no offset content; instead they have on occasion led to local assembly and production. The prospect of eventual assembly was a factor in the *Mirage*-2000 deal, and was long under negotiation for the F-16, and forms a vital part of the M-1 programme. However, with the exception of parts production for these and other systems (such as the *Falcon* 50 light transport and the *Puma* helicopter), there are no guaranteed offsets. This has meant that Egypt has been unable to exercise its option for local production, owing to lack of funds for the local defence budget or prior foreign orders.

Saudi Arabia, conversely, has used none of its massive arms deals to acquire licences for local assembly or production, but has instead made contracts conditional on the counter-purchase of offsets worth one-third of the hardware value involved. This resale content, moreover, did not have to be in material products or even defence-related, but could take the form of services or civilian products. A final implication in this context is that, although the Saudi state plays a minimal role in directing the nascent private-sector defence industry, its role in manipulating defence deals with foreign suppliers to guarantee offsets and thus provide local entrepreneurs with incentive and opportunity is major.

Organisation, History, and Activities

Saudi sources date the foundation of the industry to 1950, in the reign of King Abdul-Aziz, when the decision was taken to construct ammunition factories. The first of these is claimed to have commenced production in 1964. However, the first building block in the emergence of the modern Saudi defence industry was the ordnance factory constructed at Kharj by Heckler and Koch in the early 1980s. This formed the basis for what was loosely termed the 'Kharj Arsenal'.

In fact, the Kharj complex is the responsibility of the General Establishment for Military Industries (GEMI), formed in January 1982. Although information is lacking, it appears that the GEMI is in charge only of the ordnance factories, and not of any private companies nor joint ventures that may be set up under the offset programme. Some, though apparently not all, of the last come under the umbrella of the Military Industries Corporation, which neither owns nor directs private sector activities but co-ordinates state participation and production runs destined for the Army or security forces.

The GEMI was established in order to implement an ambitious, industrial development plan. The first version of this was a five-year plan formulated in 1971, which set as its target the construction of factories for the production of a rifle with its ammunition and of other light and medium weapons.

In the early 1980s a new plan envisaged the creation of the Prince Sultan Military City at Kharj, which would embrace five production factories, R&D centres, administration, services and support, and all that was needed to house the entire workforce with their families: housing, schools, mosques, shops, hospitals, and recreation centres.[18] In all there would be 23 administrative offices or specialised bureaux (including planning and control, for example) at Kharj. Additionally there was an industrial vocational school to provide training in mechanical, chemical, and electrical engineering to secondary school graduates, and an industrial training centre to prepare elementary school graduates in basic engineering techniques – all in order to occupy positions in the administration of the new factories.[19]

By 1986, some 75 per cent of the housing and 65 per cent of the industrial development centre was complete: the scale of the project can be gauged from the fact that one contract for housing and infrastructure was worth $266.4m.[20] The complex is reportedly capable of producing medium weapons, in addition to the light arms already in production, though no new manufacturing projects had been started by the end of the 1980s and the full workforce had not been assembled.[21]

Since the mid-1980s, the other branch of the Saudi defence industry has been developing. This is the private and joint venture sector, which is in fact very varied. An example of the private companies involved in at least some defence-related work is the KSA National Automobile Industries, which is partly owned (26 per cent) by Mercedes and which sells part of its annual production capacity of

7–8,000 trucks to the Army.[22] Another is Abdullah al-Faris & Co. for Heavy Industries, which has designed and developed an indigenous AFV. Development work on this first started in 1977, with a prototype completed in 1981. This was the *al-Jazirah*, a 6×6 wheeled APC, powered by a Mercedes diesel engine. Development of an armoured car version armed with a British Cockerill 90 mm gun turret – *Dir'-al-Jazirah* – started in 1983, and in 1987 the Saudi government decided to fund the programme. The first of 20 production vehicles was revealed in June 1989, equipped with a KHD diesel providing maximum road speeds of 105km/h and a 100 km range.[23]

Most other private companies, and especially the larger ones, are joint ventures formed under the offset arrangements with foreign companies. By the late 1980s the value of such ventures was estimated at $600m. The most important projects have been with Boeing Industrial Technology Group and General Electric (though the latter has asserted that none of its ventures are defence-related).[24] Principal examples are the $100m 'aircraft modification centre' established by Boeing at Riyadh to make alterations on fixed-wing aircraft and helicopters, and the nearby overhaul and refurbishing plant constructed by General Electric for various American-supplied aircraft and helicopter engines.[25] The former may have been responsible for unspecified modifications that Saudi officials have stated were introduced into combat aircraft, though this may have been done at Air Force workshops.[26] (Similarly, it is not clear whether the Kharj complex or Army workshops undertook other claimed modifications on such weapons as AA missiles, tanks, and light arms.)

A third joint venture is the Advanced Electronics Company (AEC), which although established under the Saudi-American offset agreement is, in fact, a partnership with British Racal. This company is to supply military tactical radio sets in kit form for local assembly, though the AEC is eventually to produce an increasing share locally, in a deal worth $300–500m.[27] Production was to take place at a new facility supposed to have been completed in Riyadh in 1990.[28] Racal is not the only British firm to operate in Saudi Arabia under the American 'Peace Shield' offset programme. Rolls-Royce is another, with a stake in the aircraft engine maintenance plant at Riyadh. An implication of this is that it will be even more difficult for the British government to find companies willing and able to invest in Saudi joint ventures under the Yamamah-2 offset agreement, as Racal and Rolls-Royce would have been prime candidates.

Assessment

In seeking to establish a modern defence industry, especially a technology-based one, Saudi Arabia can count on a small but significant number of advantages. Most prominent, of course, is the availability of funds for investment. The kingdom is, moreover, in a position, at least hypothetically, to manipulate its financial position in order to promote the development of its indigenous defence industry as a component of the massive arms deals it concludes with foreign suppliers. Given stiffening international competition among arms producers, Saudi Arabia should be able to 'play the field' and obtain the best conditions, such as licences for local assembly and production and for technology transfer.

Another advantage for Saudi Arabia – again thanks to its wealth – is its ability to train large numbers of technicians and engineers. Under the Saudi-American 'Peace Hawk' programme, for instance, Northrop Corporation trained a total of 3,000 Saudi technicians in five years (1978–83) for aircraft maintenance.[29] The $4bn air-defence deal signed with France in early 1984 also provided for the training of another 1,000 Saudi officers and technicians.[30] Although the tasks for which these personnel have been prepared are not connected with industry, they nonetheless contribute to the growing pool of technical manpower that is both familiar with advanced technology and potentially able to establish an autonomous repair and maintenance capability. In other countries these functions have become the foundation of a local aerospace industry.

Adding to the growth of Saudi technical manpower is the deliberate industrial training programme designed for implementation at the Kharj complex. Finally, Saudi Arabia can readily exploit its petroleum-based industry to produce a number of primary materials for the defence sector, such as the graphite used in the production of aircraft. Indeed, the kingdom has a well-developed chemical industry that could make a significant contribution to arms production, especially of propellants, powders, and explosives, as well as composite materials.

Despite these advantages, military industrialisation is moving slowly in Saudi Arabia. Of course, as discussed earlier, until 1990–91 the kingdom simply did not have the same incentives for rapid development in its defence industry as did Egypt and Iraq. However, there are other, more mundane reasons for this slow evolution. The Saudi industrial base remains exceedingly small, and there is as yet no industrial culture or scientific culture. Much of existing industry – both manufacturing and petroleum-based – has been provided on a turnkey basis with minimal transfer of technology or human skills. At the same time, what industry there is tends to be capital- and technology-intensive, thus reducing the opportunity to train or expand the local workforce. These disincentives also act against reducing the reliance on expatriates in industry and in military support services. These comments apply as well to the state of R&D, of which there is little or no tradition in Saudi Arabia (in common generally with most Arab countries).

The core problem here evidently relates to manpower. The industrial workforce is small, and specialists (university graduates in engineering, for example) have little experience with manufacturing or industrial culture. This is a bottleneck that cannot simply be resolved by the abundance of money. Thus training twice as many technicians or engineers will not resolve the problem; in any case Saudi Arabia may already have enough technicians, but not enough trained workers.[31] (Ironically, this may necessitate continued reliance on expatriate labour, although reducing dependence on foreigners in sensitive sectors was an aim of setting up a local defence industry.)

The desire to develop the local industrial workforce may explain, in part, the Saudi industrialisation strategy promoted in the offset arrangements. Its central objective is mainly to invite foreign investment in the establishment of local industrial companies, which may be either civilian or defence-related. To build up Saudi industry and its share in the national economy generally, and in parallel to expand the local industrial workforce and develop an industrial culture in this manner is, at best, a long-term process that is marked in the meantime by diffuseness and lack of clear aim. Specific functions may be addressed – such as maintenance or modification of aircraft at the Boeing centre in Riyadh – but defined sectors, such as aerospace, cannot be developed.

An additional problem of the current offset arrangement is that it cannot guarantee, in the absence of clear master-plans and central direction (or at least co-ordination), the transfer of technology in any field. Indeed, the terms of the offset programme had to broadened so far – to include industrial, commercial, and financial activities – that they no longer require a high technology content.[32] The problem therefore becomes even more acute in relation to the defence sector. In fact, a general comment on the prospects for Saudi military industrialisation is that it cannot proceed much further than the general state of local industry and of the economy: it will be governed by essentially the same approach as that applied in other sectors.

Naturally, the preceding is partly a result of the deliberate choice to involve the private sector in the Saudi defence industry. This makes sense in light of the shortcomings of state-dominated industries in Egypt and elsewhere, which point to the need for greater private sector involvement in the interest of increasing efficiency and innovativeness. But equally, Saudi industry is underdeveloped and the state-run military factories are limited in experience and scope; both factors call for a significant role for the state in industrialisation, in particular in the defence sector. What is needed is a balance, in which the state does more than simply ensure re-investment of foreign capital through offset deals or extend lines of credit to local entrepreneurs.

At present there is difficulty in finding Saudi or foreign private investors.[33] This may be due, on one hand, to fears about vulnerability to arbitrary decisions by the state concerning ownership, taxation, and terms of financial privileges and incentives, especially in as sensitive a sector of economic activity as defence. A second inhibition, on the other hand, is the absence of overall direction to military industrialisation. Without a clear state policy, for instance on defence exports, there is little assurance and little incentive for major investment.

An example of the resultant impasse that calls attention to the need for a balanced state role, is the four-year delay in approving a proposal by Boeing for the local assembly of a light utility helicopter, slowing down implementation of the offset programme.[34] Greater delays await the Saudi-British offset clause of Yamamah-2, which is not binding, as by the end of the 1980s the Saudi Offset Committee (headed by Prince Fahd bin Abdullah al-Saud) had not even set a timetable.[35] British investors are so reluctant, in fact, that additional concessions under consideration include reverse investment in Britain and the offer of concessionary loans, land allocations, and other assistance not normally provided in joint venture programme.[36]

These issues lead, finally, to a consideration of the impact of Saudi military spending policies on the development of the local defence industry. A major negative feature of Saudi arms deals, gradually improving with time, has been the weakness of programme design (for example, of air-defence systems), leading to inefficiency and over-spending.[37] Matters are not helped by the commission system under which contracts are granted: government middlemen have a vested interest, as much as one-fifth of contract values, in preferring to buy imported hardware and services than in developing an indigenous (possibly state-run and owned) capability.

Yet another negative feature of Saudi military contracts is that vast amounts have been allocated to foreign companies for the provision of construction and other services (maintenance and personnel support), at least some of which could be provided by Saudi or Arab firms. 'Peace Shield', for example, includes \$331m for contractor technical services and \$900m in additional construction.[38] The Yamamah-2 deal includes £1bn for construction, to be managed by

Ballast Nedam Groep (BAe-owned).[39] True, the thousands of American technicians provided under the 'Peace Sun', 'Peace Hawk', and 'Peace Shield' programmes are meant to be gradually replaced by Saudi nationals; and it is also true that local companies have put in bids for engineering and construction work; but local contribution remains low, and Saudi/Arab sub-contracting for construction and in-country support has been extremely limited.

Besides the general effect of losing a chance to spend resources locally, this pattern of expenditure has increased programme costs while limiting the opportunity to transfer more technology, if only in those particular fields. Similarly, the chance has not yet been taken to reduce spending by co-ordinating some services, such as maintenance and overhaul, with other Arab states with whom there is equipment commonality (Jordan is a case in point, as it can provide services for aircraft).

Conclusion

Despite these qualifications, Saudi Arabia remains a candidate for significant military industrialisation in the decade ahead. This is partly because it is a regional player and possesses at least a minimum of the demographic and economic requirements to become a viable producer. The experience of the Gulf crisis and war of 1990–91, added to the need to raise the local share in providing Saudi defence, may well prompt the kingdom's leadership to regard indigenous defence production as an attractive option. Conversely, there is only a limited local market for the sort of military hardware easiest to produce locally and most in demand in combat, namely combat consumables such as infantry- and medium-weapons and their ammunition and all types of ordnance.

The implication is that while Saudi Arabia can easily achieve self-sufficiency in combat consumables – and can, moreover, afford to do so at low cost-effectiveness in terms of the ratio of products to capital investment – it is precisely in all other categories of defence production that it most needs to expand. (Assuming, of course, that the objective is broader military industrialisation.) The means employed to do so at present do not appear sufficiently focused to be effective. Furthermore, the financial incentives are weak both for the private sector and even for the state, which is not in desperate need to make savings or increase exports.

Nonetheless, a long-term view shows the importance of diversifying economically and of increasing self-reliance in relative terms, in order to ensure domestic stability and reduce dependence on high-tech, high-cost weapons, and so military industrialisation may remain a goal. However, the driving force facilitating this process and shaping it will be the availability of surplus capital, which will continue to substitute for a proper industrial base, adequate workforce, and comprehensive defence industry strategy for the foreseeable future.

7

Other Arab Countries

Egypt, Iraq, and Saudi Arabia are effectively the only Arab countries with an existing or emerging defence industry. Of the others, Kuwait has explicitly rejected the idea of developing a local industry: parliament expressed its opposition in spring 1984, and was immediately followed by the Defence Minister, who announced that the GCC was forming a plan for joint military industrialisation.[1] Despite occasional statements since, however, the GCC has not in fact produced, let alone implemented, a definitive, joint plan.

Furthermore, despite much furore in the Western press concerning missile tests by the German company Otrag (1979–81) and the construction of purported chemical weapons plants at Rabta and Sirte, Libya too has not developed significant industrial capabilities in the defence sector.[2] It does, however, have a Military Industrial Organisation, within which the Technical Industrial Corporation would probably be responsible for special areas of research, such as missiles and chemical warfare, should they be attempted. The latter body is therefore probably in charge of the *al-Fateh* (*Conqueror*) ballistic missile programme: this was reportedly being conducted in the late 1980s with the assistance of designs and components (and possibly experts) supplied by West German companies, though there was no further evidence of this effort by mid-1991.[3]

Nonetheless, a few other Arab states have minimal manufacturing capabilities, or else have sought to establish their own military production facilities. These are Syria and Algeria, and to a significantly lesser degree Jordan – all of which are described below. Not included in this review are Morocco, Tunisia, and Sudan, each of which is claimed to have produced small arms at one time or another, though there is no indication of such activity at present.[4]

Syria

In Syria, the Establishment for Defence Factories has for about two decades been producing ammunition for small arms, and possibly also for machine guns and such medium weapons as mortars. A proposal to produce light mortars and RPG-7 anti-tank rocket launchers jointly with the PLO was also mooted in 1974 but subsequently discarded.[5] Israeli sources have suggested that nerve agents are being produced at Syrian plants near Lattakieh, in parallel with design work on suitable chemical warheads for use on tactical ballistic missiles, but this is not confirmed. According to these and Western sources, Syrian production levels are nearly as high as Iraq's, which were estimated to be 60 tons of mustard gas and 4 tons each of tabun and sarin.[6] The one weapon that Syria is known to have produced is a self-propelled 122-mm howitzer. This is the Soviet D-30 mounted on the rear of a T-34 tank hull, providing a theoretical 360° traverse (120° in practice) and supplied with 40 rounds packed in containers fixed to the external hull sides.[7] From their design these SP guns – of which only a few dozen at most are in service – may have been produced at Army workshops rather than in the defence factories.

Given the extent of Syrian defence expenditure and the volume of military imports, a degree of local industrialisation would seem sensible, especially of combat consumables. The country, moreover, has established (albeit minor) engineering and chemical industries, as well as modest reserves of some primary materials such as titanium minerals and oil. However, it may be that Syria, as with Egypt in the 1960s, finds it cheaper to purchase ready-made Soviet hardware than to produce it locally. Indeed, it is unlikely to achieve better economies of production in the absence of Arab financial backing or industrial co-operation. The imperatives of Syrian defence, national security, and foreign policy seem to push in the same direction. Faced with its heavy involvement in Lebanon and the continued likelihood of confrontation with Israel on the one hand, but assured of uninterrupted Soviet supplies of most military items on the other (until 1991), Syria appears to have opted for reliance on its one secure source of armament rather than embarking on a costly course of local industrialisation.[8]

Algeria

Algeria has almost no military production at present. Its main activity in the past was boat-building. The state-owned Chantier Naval shipyard at Mers el-Kebir has produced *Kebir*-class fast patrol boats under licence from the American Brooke Marine: the contract covered seven boats, four of which were constructed in 1985–86, and three in 1987.[9] The same yard also produced 12 18-ton fishery protection boats in the early 1980s, as well as a number of landing craft.[10] There has been no evident Algerian military production in any other branch of hardware. What is known is that an MoU was signed with Britain in December 1983, covering co-production of some of a variety of ammunition, infantry weapons, AFVs, naval vessels, and aircraft on offer for sale.[11] The *Hawk* trainer/close support aircraft was one item under consideration, as was the *Alpha Jet* in later years. This was not taken up, however, probably for financial reasons. Nor were subsequent requests for bids from Britain, France, and Germany for help in setting up facilities for the production of ammunition for 12.7-mm machine-guns, 23-mm automatic cannons, and 60/82-mm mortars, 82-mm mortars, ATGMs, light vehicles and AFVs, and electronics.[12]

The main incentive for military industrialisation from Algeria's point of view is probably a mixture of national security considerations and economic calculations. Although the possibility of large-scale war with neighbouring Morocco is not high, tension has dominated relations for two and a half decades. Algeria is, moreover, seeking to diversify its arms sources, in order to ensure greater independence. Economically, the establishment of a defence industry would conceivably lead to eventual hard currency savings and encourage the growth of local manufacturing capabilities and manpower, which saw a decline in the second half of the 1970s, and so employ oil revenues in more effective development. This would also make better use of the chemical, iron and steel, and heavy mechanical and electrical engineering industries that Algerian industrial strategy has focused on.[13] Among the last are an automobile assembly industry, which in other countries has provided a basis for defence production. Given the lack of an urgent national security imperative, production economies might best be achieved by forming a joint defence industry and spreading the costs with Morocco (especially), and potentially with Tunisia and Libya, all of which are members of the Maghreb Arab Union.

Jordan

Compared with Syria and Algeria, both of which have a significantly larger economic and demographic base, Jordan seems an unlikely candidate for military industrialisation. But the kingdom has considered establishing a modest productive capacity. One proposal discussed with Britain in the early 1980s envisaged local manufacture of a number of auxiliary items, such as vehicle batteries, sights, tyres, and spare parts. A second idea was to develop an indigenous electronics industry, with an eye on the Singaporean model, though British experts have suggested that the country should concentrate on computer software instead.

One project that has been implemented, however, is the updating of the ageing British *Centurion* tanks in the Jordanian inventory. Virtually every component has been modified or replaced, including suspension, engine, transmission, turret components, and main armament, while full fire control and optics have been added.[14] The design is Jordanian, though the kit is imported, and can be used to upgrade M-48s and M-60s. At a unit cost of $80,000 in 1983 (compared with the $600,000 it took to buy a new M-60 at the time), the converted *Centurion* – known as *Tariq* – is generally superior to the M-60A3.

In the late 1980s, the elements of a new military industrial strategy appeared. The central component of this development was the formation in July 1988 of the Jordanian Technology Group (JTG), which is a government-funded, industrial development organisation. JTG has already started a number of local or joint ventures to develop certain types of equipment or produce them under licence. Few of these involve combat hardware, and several revolve around services or non-combat support equipment. JTG aims ultimately to assist the transfer of Western military technology to the Arab region, especially through high-tech production activities.[15]

Underlying this move is the view expressed by Crown Prince Hasan of Jordan that Arab military acquisitions reflect a lack of co-ordination, resulting in duplication of effort and expense and loss of opportunities to standardise or economise on equipment.[16] His evaluation of Jordan's particular role is that it should utilise its skilled manpower in order to provide the spare parts and maintenance for the high-tech hardware in Arab service. An example of this approach would be to to send Jordanian engineering teams to the states concerned, in co-ordination with the original manufacturers. A core objective would be to achieve greater savings in defence spending generally and reduce unemployment, increase exports, substitute for some imports, earn hard currency, and acquire technology.[17] As part of this process, the Jordanian government has also introduced an offset clause into contracts over a specified value, with the ultimate aim of obtaining a 60 per cent level of returns on all such purchases.

Among the new ventures launched in Jordan since 1988 is the MoU signed between JTG and British Thorn EMI Electronics (Electro-Optics Division) for the local licensed production of hand-held thermal imaging equipment, possibly from 1990.[18] A second MoU has also been concluded with the French Sopelem, under which the latter company will provide 40 sets of helicopter night-flying goggles, while the former establishes a facility for maintenance of night vision equipment.[19] A more ambitious venture, already under-way, is the local assembly of the privately-designed American

Schweizer Model 330 turbine training helicopter. This will be produced by a joint venture (Jordan Aerospace) combining JTG with Schweizer Aircraft Corporation and Cypriot Aerospace Marketing Services, with the first helicopter appearing in late 1990. In addition to increasing local content steadily in the expected production run of 100 S-330s, Jordan Aerospace will also undertake the refurbishing of old helicopters.[20]

In parallel with the last programme, JTG plans to expand Jordan's existing civilian engine overhaul facility, in order also to test and maintain military jet engines. Primary candidates are the Snecma *Atar* 9K-50 for the *Mirage* F-1 and eventually the M53-P2 of the *Mirage*-2000.[21] Other planned projects include the manufacture of tyres for both military and civilian vehicles, development of a fast attack vehicle mounting a machine gun or anti-tank weapon for desert use (based on a 'buggy' with a Volkswagen engine), construction of facilities for testing and maintaining military aircraft engines, and production of several calibres of ammunition at a new plant.[22] Both the last project and the aircraft engine overhaul facility have previously been proposed or even initiated in Jordan, though with limited success, partly owing to lack of Arab response. (The new Saudi aircraft modification and engine maintenance centres being developed under its offset agreements with American and British companies will obviously compete for major markets, as will Egyptian ammunition factories.)

In conclusion, although Jordanian industry is relatively undeveloped, its current plans appear well suited to its particular capabilities on the one hand, and to Arab needs on the other. The formation of the Arab Council for Co-operation in February 1989 appeared to offer a convenient framework both for inter-Arab co-operation and for marketing goods and services, though its collapse during the Gulf crisis cast serious doubts on the future of Jordanian military industrialisation. In all cases the kingdom's main attraction is the availability of technical manpower (though there is little industrial experience specifically), which it can either export to undertake technical follow-on support for military hardware or employ in local high-tech ventures such as computers and electronics.

The PLO

Finally to be mentioned are the military productive efforts of the PLO. Although the organization lacks the full resources of a state, it nonetheless presented a modest example of how local initiative and dynamism could be usefully employed. The channel for indigenous efforts was the 'Scientific Committee' of Fateh, which was formally established in the early 1970s, after commencing operation in

1969.[23] It maintained several workshops and maintenance centres in Lebanon until 1982, as well as a small number of productive facilities in Iraq until 1974.

One of Fateh's earliest products was a copy of the RPG-7 anti-tank rocket launcher, which remained in production from 1969 to 1982. Also manufactured locally were various types of explosives, and fibreglass boats. The Scientific Committee also oversaw the addition of armoured plating to Unimog trucks, to make APCs, and also armoured three mobile radar trucks for Fateh forces. As significant was local design and development work, which led to the production of copies of Soviet-designed 122-mm rocket launcher tubes and their firing mechanisms, that were then combined in multiple banks and mounted on vehicles. Repeated attempts were made to extend the range of the 122-mm artillery rockets themselves, though with little success. Fateh also produced wholly indigenous electronic detonators, and commenced work on other electronic equipment before 1982.

Ultimately, these efforts did not represent a substantial productive capability, something which required greater political decisiveness, funds, and technical abilities than were available. Work progressed largely on a trial-and-error basis, rather than on systematic scientific and theoretical foundations. Nonetheless, it was the initiative shown and the attempt to utilise local resources that made the PLO programme interesting. Indeed, the organisation's dynamism led Syria, Libya, and Algeria to suggest joint efforts on various occasions in the 1970s and 1980s. Thus, although a very minor industrial actor, the PLO offered a microcosm of some of the shortcomings and potential of Arab military industrialisation.

PART THREE

Technical, Operational, and Policy Assessment

8

Technical and Operational Assessment of Arab Military Industry

Having reviewed the history, organisation, and activities of the three main Arab defence industries in detail, it becomes possible to assess their technical and operational performance. The relevant technical indicators include the distribution of military industrial activities according to productive mode (assembly, manufacture, and parts production); origin (licensed copy, reverse-engineering, modification or new design); and technological level (high, low, or intermediate). The operational dimensions relate to the structure of the arms industry (private/public sector role) and other determinants (funding and manpower), and to its share in civilian markets and exports.

Productive Modes

An immediate and expected observation when reviewing Arab defence industry programmes is the broad correlation between productive mode and technological level of the product. Items at the higher end of the technological scale, such as aircraft and AFVs, tend to be assembled, whereas low-technology hardware, such as infantry weapons and ammunition, are often manufactured entirely locally. As the technological level of an intended product rises, complete manufacture gives way to production of parts and components, which are used in the assembly of larger systems.

Assembly

The prominence of assembly as a productive mode in Arab defence industries is most obvious in the aerospace branch. Of seven programmes involving production of a complete aircraft (fixed-wing or helicopter) undertaken since 1950 (all in fact in Egypt), all but one have started with local assembly, either partial or full. The short-lived HA-300 fighter in the early 1960s is the only exception, as it was being developed from Spanish designs and prototypes and had not yet entered production anywhere. Of the other six aircraft, assembly gave way to production in three cases: *Gumhuriyyah*, HA-200, and *Tucano* – but not for the *Alpha Jet* and F-7 aircraft or the *Gazelle* helicopter. Additionally, the two major aerospace projects initialled by Egypt but not implemented – the *Lynx* helicopter and the *Mirage*-2000 fighter – also involved assembly (as well as manufacture of some parts) for their full planned production runs.

Notably, the shift from assembly to production does not imply local manufacture of the complete aircraft, with all or most of its components and parts. Rather it refers mainly to manufacture of the airframe and a number of minor parts, but not of such major systems as the engine, avionics suite, and electronics, which are purchased from abroad and incorporated in the final assembly of the aircraft. As a senior official in the Egyptian aerospace industry explained, 'Aircraft contain many operating systems . . . no factory in the world produces them all, instead it buys them from [other] specialised factories . . . So when we refer to local production, the comparison is with the amount of production at the mother company making the [same] aircraft.'[1]

So when the AOI claims to be producing 73 per cent of the *Tucano* locally in Egypt, this is not a proportion of the entire contents and cost of the aircraft, as it excludes virtually all major components and sub-systems, which have to be purchased from abroad. The latter may represent a 70 per cent share of final cost in an aircraft such as the *Mirage*-2000 (though less in a primary trainer like the *Tucano*).[2] What the AOI's Qader Factory does on the *Tucano* is 73 per cent of what the equivalent division of Embraer does in Brazil (leaving aside design and development); the main difference is that Qader then imports the remaining components, whereas Embraer may buy a greater part of its needs locally from other Brazilian companies. The same applies to the Dassault-Breguet/Dornier *Alpha Jet*, which is 15 per cent produced in Egypt: other French and German (and European and American) firms provide many components besides the airframe, whereas the Helwan Aircraft Factory has to import them. For the AICs (or producers such as China), therefore, the overall or *national* share of production is much higher than the Arab defence industry as a whole can achieve. Evidently this implies for the Arab countries that the savings to be made nationally through the use of cheaper local labour are limited by the size of the share of total production, and that technology transfer will also be limited to certain areas.

Much the same logic applies to local assembly and production of AFVs, especially tanks. The latter vehicles are among the most difficult weapons systems to produce, largely owing to the need to manufacture large foundry pieces.[3] In addition, AFVs may contain special armour, and in all cases require relatively advanced technology for their engines, transmissions, suspensions, and electronics. Thus the proposed M-1 tank programme for Egypt involves only a 19 per cent local workshare, and does not include engine, transmission, armour, or fire control system (FCS).[4] And it is probable, though no definite information was made public, that Iraq would not have manufactured several major components of the T-72 even when full production was reached, according to its contract with the USSR.[5]

Finally, the production of parts offers one, partial solution for Arab defence industry. The Egyptian aerospace industry manufactures a number of parts for the aircraft it assembles and produces, and even acts as a sub-contractor to the original suppliers by re-exporting parts for their production lines. (Producing some parts for an aircraft is not equivalent to producing it, though.) So far the items concerned tend to be airframe or simple engine components that involve milling and steel-work. However, it may be possible to expand this line of activity and so gradually increase the local workshare and acquire more technology, by producing more advanced components. A partial, gradual approach would be manageable technically, and more likely to be accepted by the original designers and suppliers. This is most evident in the electronics industry, in which Arab factories rely predominantly on assembly rather than manufacture. Yet the sector supplies most weapon systems, and so any advance in it would widen the local contribution to other fields.

Manufacture

Local manufacture includes a variety of industrial activities. One is the production under licence of imported designs, while another involves unlicensed production of copies of originals through reverse-engineering. Alternatively, completely original designs may be developed locally, or else partly-new weapons are produced by introducing certain modifications on imported designs that are then produced locally as integral systems. In every case, the actual proportion of local manufacture – the extent to which every component part or primary material is physically shaped and finished within the country – may vary, but is assumed to be high (as an arbitrary measure, over 50 per cent of total production cost).

In the Arab case, the vast majority of locally-manufactured hardware – arms and munitions – is based on foreign designs rather than local ones. Most of the products falling within this category are infantry and medium weapons, ranging from 7.62-mm AK-47 or G-3 rifles, through RPG-7 rocket launchers and 60/81-mm mortars, to 14.5/23-mm AA guns and *Swingfire* ATGMs. A number of heavier systems are also manufactured, such as 122-mm and 130-mm guns, 120/160-mm mortars, 122-mm MLRS, and Scud-B ballistic missiles (and the last's derivatives). The distribution of manufactured systems according to the source of design – transferred blueprints for licensed production, or reverse-engineered copies – appears to be even. Many, but not all, of the Soviet-designed arms and munitions produced by Egypt are reverse-engineered, whereas comparable items produced by Iraq were licensed. (Of course, the licenser is often not the original designer but a secondary producer, so that Iraqi copies of the AK-47 rifle and 122-mm D-30 gun were licensed by Yugoslavia, the RPG-7 by China.) It is this wide range of combat consumables that forms the mainstay of Arab defence industries and their principal contribution to the military needs of their countries.

A second category of locally manufactured armaments is made up of original, non-imported designs. Most of these are concentrated in the field of rocket artillery, although the Egyptian *Fahd* and Saudi *Jazirah* APCS and the Iraqi *Yamamah* RPV are additional examples. The obvious implication of the prominence of rocket systems is that Arab ability to design, produce, and test the necessary components and associated tooling and materials for other weapons categories – such as gun artillery – is limited. Rocket casings and propellants are easier to make than those for guns, and their launchers are also simpler than gun carriages and tubes. Furthermore, because the inaccuracy of rockets is an accepted element of their operation (especially compared with guns), imprecision in manufacture can be accommodated. That said, some of the new Arab rocket systems are also based on older imported designs – as is the case with the Iraqi *Husein* and *Abbas* ballistic missiles derived from the *Scud*-B – though several more MRLs are completely original: the Egyptian VAP and D-3000/6000 80mm, and *Saqr*-80 and the Iraqi *Sajeel* 81mm, *Nasr* 240mm, and *Ababil* 400mm.

Other significant rocket developments at the end of the 1980s were the *Badr*-2000 and *Walid* ballistic missiles being developed by Egypt and Iraq, respectively. These may have preluded more ambitious space and nuclear programmes originally planned for the 1990s. As important was the *Faw* ATBM, which was claimed to enjoy post-launch guidance. The new family of tv- and laser-guided air bombs and the *Sumer* mines also supposedly involved guidance technology. However, it is not clear how many of the crucial components were produced (let alone designed) locally; the final systems may have been new but still consisted of imported parts. This possibility is a feature of some of the more advanced systems produced by Arab industries. The *Fahd* APC is an Egyptian design, but incorporates a German engine, transmission, and chassis. Evidently, as the product rises higher on the technological scale, the proportion of local manufacture (and local design) drops.

Partial Manufacture and Modification

Lying between assembly and part production on the one hand, and complete manufacture on the other, are a number of defence products that are based to a considerable degree on imported designs and manufacture, but also contain a sufficient share of local labour or technical modification as to result in a distinct end-product. Obvious examples are the 122-mm artillery rocket, the *Saqr-Eye* man-portable SAM, and the *Amun* AD system, produced by Egypt, or the 125-mm-gunned T-55 tank, the 155-mm M-46 and the 210-mm SP *Fao* gun, and the *Barag*-guided SSM produced by Iraq. In each of these items, the Arab defence industry concerned has modified a few components in order to extend range, increase accuracy and lethality, prolong useful life, or even alter function. In most of these cases too, the modifications are implemented locally, but largely on imported hardware. Thus the Iraqi T-55 couples a T-72 gun with the old hull and adapted turret, while the Egyptian *Amun* combines Swiss 35-mm AA guns and radar control with American-made *Sparrow* SAMs.

Just how much local input was contributed to each of these products is difficult to assess, and so defining this category of Arab arms production is not easy. Where, for instance, is the dividing line between a *Tucano* trainer in which a substantial proportion of the airframe and even engine is being manufactured locally (though well under 50 per cent of the total aircraft cost), or an *October*-class FAC/M (originally a licensed copy of the Soviet *Komar*) in which the hull and fittings are Egyptian-made, and the *Amun*, which contains 14 local modifications but remains 95 per cent foreign-made? And why consider the *Fahd* APC as manufactured rather than assembled in Egypt? At issue here is the question of local input to design and technology, which is addressed next.

Technological and Design Levels

As with the distribution of productive modes between assembly and manufacture, a somewhat similar situation applies to foreign and local roles in military design and development, in that local design input tends to be at the lower-to-intermediate levels of production, tailing off almost completely at higher levels. However, the similarity ends there: just because local design input is at the lower end of the technological scale and cannot cope with advanced weapon systems (aircraft, for example), this does not imply a marked ability to design complete basic weapons (assault rifles, for example) instead. That is, local design ability is concentrated overwhelmingly on modifying any existing system through partial design changes, or through altered combinations of existing components.

In the preceding section some emphasis was placed on defence products produced wholly or partially within the Arab countries. Much of such local production depended on imported designs, which were copied under licence or through reverse-engineering – the former method may have provided more know-how, but the latter stimulated indigenous research – while a significant proportion of local products resulted from new design work. But just how much local input to such R&D there has been, as opposed to foreign assistance, is a doubtful matter. Thus whether in reverse-engineering an established, imported item (weapon or munition), producing an entirely or substantially new one, or modifying an existing system, Arab defence industries appear to have relied heavily on foreign design input at all stages. Although it runs somewhat counter to the thrust (both implicit and explicit) of official statements, the conclusion here is that many of the advances credited to local defence industries would not have been possible without such external assistance. This is not to decry Arab efforts and achievements, but to present a more realistic understanding of the current situation and the technical difficulties to be surmounted on the path to broader military industrialisation.

A brief review of certain weapon systems manufactured or modified by Arab defence industries supports this assessment. In Egypt, the D-30 122-mm howitzer was reverse-engineered, but with Chinese help in barrel manufacture. The 122-mm artillery rocket was modified to extend its range and carry a new cluster warhead, but this was done with the aid of French and German sub-contractors. Unlike Egypt, Iraq tended to produce many of its combat consumables under licence, possibly spurred by the urgency of its needs for the war against Iran. However, it too relied on foreign assistance in designing new or hybrid weapons. This was the case with the 210-mm *Fao* SP gun, which fielded an entirely new calibre apparently developed with French help.

This weapon also raises another aspect of the design issue, namely the distinction between product and system design. In the items mentioned so far, foreign help aimed at developing specific hardware components. Of equal importance, though, is the external role in designing systems according to Arab technical specifications or combat requirements. Cases in point are the Egyptian T-54/55 conversion (*Ramses*), 122-mm SP gun, and *Nile*-23 and *Amun* AD systems. In Iraq, the same applied in the 1980s to the 155-mm *Majnoon* and 210-mm *Fao* SP guns and the *Husein/Abbas* ballistic missiles (the latter with North Korean and Egyptian help, but possibly also West German) though the design status of the M-46 155-mm gun and T-55 125-mm tank conversions (among others) is unclear. In all these cases, noticeably, the end-product consists largely of imported components and sub-systems, which are only assembled or part-produced locally.

There has been appreciable local contribution nonetheless, whether totally independent or in collaboration with (but not subordinate to) foreign assistance. Examples are the modernisation undertaken by the Egyptian Army of the T-55 tank in its workshops and the uparmouring of the M-113 APC, or the Iraqi uparmouring of the T-55. And according to Egyptian officials, the electronic guidance systems for the *Saqr-Eye* (SAM-7) and *Early Bird* (SAM-2) were developed locally through reverse-engineering and modification. However, the greater Arab input has been in co-operation with foreign designers and technicians; this is a known fact for the large family of tactical ballistic missiles under production or development, and is also probably the case for a wide variety of new Iraqi products, from the *Baghdad*-1/*Adnan*-1 EW aircraft to the *Majnoon* and the *Fao* SP gun and the *Sumer* naval mines.

That said, most design modifications implemented by Arab defence industries – regardless of the original source of the R&D and technology involved – have divided broadly between three main categories: system design, rocket artillery (of all calibres), and electronics. In the first case, system design modifications refer to the alterations introduced to existing weapon systems in order to improve performance or alter function. This includes the 122-mm SP gun proposed by the British ROF and the American BMY for the Egyptian Army, the last's *Ramses* (T-54) and T-55 tank upgrades, and its *Amun* and *Nile*-23 AD systems, and Iraq's own two T-55 upgrades, An-12/I1–76 airborner tanker and *Baghdad*-1/*Adnan*-1 EW conversions, 155mm M-46 re-barrelling, and 155/210-mm *Majnoon/Fao* SP guns and 120-mm single/quadruple-barrel mortar carriers. In each case little or no modification has been done on the individual components themselves – from engine, and hull to main armament and guidance – but they have been brought together in new combinations. Of course, at times such modification involves little more than the addition of a few ancillary systems, such as laser range-finders, new fire control systems (FCS), and smoke dischargers to tank hulls and turrets, in which case the added items are often imported, and reflect only limited local input (of either design or production). And as discussed earlier, much of this system design modification effort is undertaken by, or with, foreign assistance.

Other design modifications have been in rocket artillery, and include the development of entirely new weapons, such as the 80mm VAP and D-3000/6000, 81-mm *Sajeel*, 400-mm *Ababil*, and the *Saqr*-80, *Badr*-2000, and *Waleed*. Leaving these aside, though, most partial alterations have centred on the warhead and propulsion system. Thus both Egypt and Iraq have developed new 122-mm cluster warheads, and extended the rocket's range. Similarly, the *Frog*-7 and *Scud*-B derivatives (*Husein* and *Abbas*) have benefited from modifications to their warheads and rocket motors. Seen in the light of the nature of system modifications, and of the reliance on imported components or designs, the concentration on rocket artillery indicates the imbalance in Arab capabilities. This is both confirmed and countered by the fact that some Arab initiative has taken place in the field of electronics. Egyptian engineers have copied some SAM guidance systems, and co-operated with foreign firms in designing new electronics packages (including IR sensors and IFF, for example) for them. They have also introduced 14 modifications to the *Skyguard* AD system to convert it into *Amun*, and helped to design a strategic EW system for the country's air defence. Iraq has integrated European electronic and radar equipment into its *Baghdad*-*I*/*Adnan*-1 EW aircraft, and developed an interface to allow the Soviet As-14 laser-guided ASM to be fired from French *Mirage* F-1s.

Although not immediately obvious, an implication of the review so far is that in all but the most isolated cases Arab design and productive efforts are focused solely on combat equipment and consumables, to the near-total exclusion of ancillary and supporting equipment such as mine-ploughs, engineering and armoured recovery vehicles, gun tractors, tank transporters, bridge-layers, trucks and bulldozers, and dozens of other non-combat systems. These items are essential to the functioning of any modern army, and are imported in massive quantities. Yet beyond a mine-detector and a rocket-assisted infantry mine-clearing system, diesel units, field bakeries, and NBC protective clothing developed in Egypt, there has effectively been no Arab investment in this sector.

Secondly, the Arab defence industries of the present are a far cry from the ambitious programmes of the early 1960s, when Egypt sought to develop its own indigenous combat aircraft and ballistic missiles. The state of military technology then was such that the gap between LDCs and AICs was not unbridgeable, and Third World countries could hope to sustain the costs of designing as well as manufacturing new equipment. However, the rapid pace of technological innovation has left the LDCs ever further behind, especially as they have had to invest growing resources in development, and even in ensuring mere survival (or paying off debts). To approach an indigenous design capability once more is still a remote prospect, especially in aerospace, though Brazilian success in producing AFVs and Arab ability to develop ballistic missiles in collaboration with other LDCs suggests that progress can be made.

The third point, not unrelated, is that the issue of design and technological level affects manufacturing techniques and know-how as well as the end-products. For many years even relatively simple items, such as artillery shell casings, had to be imported for filling at local factories. An associated shortcoming is the fact that many productive facilities have been imported and supplied on a turnkey basis, greatly reducing the amount of technology and skill transferred. Thus manufacturing know-how remains a bottleneck in Arab defence industries, as does the high degree of dependence on external sources for tooling – all of which impede the accumulation of technology and limit the ability to undertake innovative design and development.

Operational Dimensions and Determinants

Five major operational factors and determinants affect the performance and level of development of modern defence industries: their relationship to civilian industry; role of the private sector; source of funds; availability and skill of technical manpower; and volume and importance of exports. Each of these has a direct bearing on the way the defence industry is run, its impact on the general economy, its ability to acquire technology, and its long-term viability.

Relationship to Civilian Industry

The situation with regard to the relationship between civilian and military industries varies widely from one Arab arms producer to the next. Of the three major countries concerned, only Egypt has allocated a substantial part of its defence production capacity to manufacturing civilian goods. At the opposite end, until 1990, was Iraq, with no definite link known to exist, though certain plants may produce materials (such as chemicals or steel) for both the military and the civilian sector. On the other hand, although Saudi Arabia produces no civilian goods at its ordnance factories, it is using the offset clauses within its foreign arms purchases to expand all branches of its civilian economy as well as expand the local defence industry. Of course the reverse relationship is also important, namely the contribution made by the civilian industry to the defence sector.

In Egypt, both the NODP and the AOI produce civilian goods, and have allocated an estimated 45–50 per cent of their productive capacity for the purpose, according to Egyptian figures and worth some $350–400m annually.[6] A semi-official list of Egyptian civilian products includes electrical and industrial machinery, household appliances, metal products, batteries, tools, and agricultural tractors.[7] Notably, these civilian production lines were not in operation before the establishment of military facilities. Rather, defence production capacity was diverted to civilian activities from 1958 onwards, in order to solve the growing crisis of the defence industry and utilise excessive capacity. Implied here is that civilian production was not a natural extension of military sector development or of defence-oriented R&D. Indeed, although a quick survey of Egypt's civilian products may show several that can be produced in conjunction with military goods, not one specifically owes its existence to a military programme. While civilian production has been used to improve the economics of the defence industry, it has neither taken on a life of its own nor been stimulated by the latter.

The weakness of civilian-military industrial interaction was very evident in the case of Iraq up to 1990. There too, certain civilian plants may have contributed to, or benefited from, specific military programmes such as chemicals and steel production. Iraqi officials have pointed to the involvement of some defence factories in producing ordinary civilian goods. But once again there is no evidence that either type of product – civilian and military – owes its existence to the other, nor that there is a dynamic linkage (in either direction) between the defence industry and the civilian economy. At most, the Iraqi Military Industrialisation Authority has undertaken civil construction projects (bridges, dams, and grain silos), but this is a function of engineering corps in many armies and hardly dependent on the special abilities of the defence industry. Indeed, among the civilian enterprises cited by Jordanian officials as evidence of the civilian-military economic linkage is the Army's role in laying pipelines and building bridges.[8]

In the case of Saudi Arabia, the state-owned ordnance factories undertake no civilian production. The situation is different for the small number of private companies partially involved in military production, as the bulk of their activity is in fact directed towards the civilian market. Once again, though, no necessary relationship between the two spheres of activity has emerged, nor has the civilian sector benefited from military R&D. As mentioned above, the Saudi government is attempting to expand its industrial base and the economy generally (services and commerce) by utilising offsets deals acquired through military contracts with foreign arms suppliers. Yet here the linkage is remote and cannot be regarded as a civilian-military linkage, except in those ventures that explicitly serve both sectors. The obvious examples are the Aircraft Modification Centre and Engine Overhaul Facility, both of which provide technical services for the kingdom's military and civilian aircraft.

On the reverse side of the coin, Arab civilian industry is making a limited contribution to the defence sector. The most obvious input is provision of steel and metal products, plastics and rubber, and chemicals. Official data are rarely available, but the general impression is that the actual production of specialised primary materials for military production, including types not normally used in civilian goods (such as tritium, carbon-carbon, and hardened steel, for example), is considerably lower than statistics of production volumes would indicate. Besides, the civilian sector in all Arab countries is in need of more primary products than can be manufactured locally, and makes up the deficit through imports. Theoretically, defence industry demand should spur expansion in the feeder civilian industries, and several such projects are planned in Egypt and Iraq, but actual implementation is dependent on a number of factors, among which are endemic problems and systemic determinants in the national economies concerned. This naturally affects other civilian products for the military market, such as tyres and automotive parts, the quantity and quality of which are constrained by the general level of the industry.

With these qualifications in mind, the Egyptian civilian industry is evidently the one most capable of feeding the military sector, compared with Iraq – even before the 1991 Gulf war – and Saudi Arabia. This is owing to its long experience, and to the gradual expansion of the private sector in specialised fields, both elements lacking in the Iraqi and the Saudi case. However, Saudi Arabia possesses a large petrochemicals industry, and so is well placed to feed both its nascent civilian and military manufacturing industries, while Iraq may still have the potential to develop both heavy and petrochemical industries.

But since the performance and capabilities of heavy industry are closely tied to the structure and capabilities of the national economy, the eventual contribution of the civilian sector to Arab defence industries may well be more effective in such specialised areas as electronics, computer sciences, and optronics. This is a role played partly by a handful of Egyptian private companies at present (including foreign joint-ventures and subsidiaries, which plan to divert part of their automotive and electronics lines to producing military goods), possibly to be followed by the Racal subsidiary in Saudi Arabia. (Before the 1991 war, the Saad-13 factory in Iraq was another prospective contributor.) Their contribution remains woefully inadequate so far, but an option for the future is the emergence of dynamic companies devoting substantial efforts to R&D and producing goods for both civilian and military markets.

Private Sector Role

Evident in the preceding section is the potential importance of the role of the private sector, which, however, has a particularly weak input to Arab defence industries. Compared with the record of public sector performance, private participation can achieve greater efficiency and rationalisation within productive facilities, raise accountability and tighten standards, attract qualified personnel and investors, and develop feeder industries. However, among its main drawbacks – in contrast to state-owned enterprises – are its lack of the volume of capital required to establish military industries and its dependence on expanding sales and exports for survival.

The issue of the private sector's role divides into two: direct private participation in defence production, and indirect contribution through civilian feeder industries.

As seen in the country case-studies, private defence companies are few, and represent only a fraction of total military production. Their share is highest in Egypt, where half a dozen such firms earned an estimated $60m in 1987, or 15 per cent of all military products sales (though this figure may actually have included an appreciable proportion of items such as tyres and metal parts, or non-manufacturing activities such as construction).[9] In Iraq, the private defence contribution was effectively non-existent. Saudi Arabia, on the other hand, is actively using its offset programme to establish private defence companies (such as the Racal subsidiary) which receive government SIDF loans, in order to build a predominantly private defence industry (excluding the state's ordnance factories).

At present, private Arab defence companies are most active in the fields of electronics, computers, and optronics, which are generally among the least developed branches of the state sector. Two probable reasons for this are the ability of private firms to attract highly-qualified personnel and to enter into profitable partnerships with foreign companies that can offer advanced technology. An added attraction, especially for Arab states with limited funds, is that setting up such specialised industries requires far less starting capital than do major weapons facilities. This is evident in Jordan, where the Jordanian Technology Group (JTG) was founded in 1989 to undertake and co-ordinate high-tech productive projects with local private and foreign investors.

Furthermore, private participation can reach major proportions, with the Egyptian (semi-private) Arab International Optronics enjoying a turnover of $25m in 1986, while the Saudi Racal subsidiary is now undertaking a $300m programme to produce military radios. The Saudi offset programme should eventually expand the private share in the defence industry even further, though this is by no means guaranteed, and much of the investment might go to other sectors. Moreover, few Saudi private companies are opting to diversify their civilian activities to take up military production as well, effectively leaving the field to foreign joint ventures and subsidiaries.

In contrast to the direct private role in Arab military industrialisation, there is the broader issue of indirect participation through civilian feeder industries. However, in the Arab countries concerned, the principal feeder industries are state-owned, with only a minor private share. But if the Arab governments are not to take on an even greater financial burden in establishing and running the feeder industries needed by local defence sectors, the private role must expand. Moreover, private participation, especially in the associated civilian industry, is important to ensure improved economies of production and management. This is even more the case of companies that supply secondary items (machine parts, tyres, and other rubber products), rather than heavy primary materials (steel and chemicals), as a sideline in addition to their activity aimed at local civilian and export markets. Seen the other way round, greater private participation could improve efficiency at defence factories diversifying into civilian production.

However, the private sector suffers from major disincentives to participation in Arab defence industries, whether direct or indirect. A primary one is the shortage of funds: military production requires heavy initial outlays, and the cost of acquiring and reproducing major weapon systems or other advanced ancillary equipment is prohibitive. This is particularly so if production lines can only be directed to the manufacture of military goods, and do not allow additional revenue from civilian activity. So considerable financial backing from the state is necessary for private companies to launch defence projects. This is the specific purpose of Saudi Arabia's SIDF, but is notably lacking in Egypt and, of course, Iraq (pre–1991). Nor are other traditional sources of funding readily available, such as commercial bank loans, and for a variety of reasons, among which is fear of incurring major losses in an uncertain sector. Another reason, itself a distinct disincentive, is the role of government bureaucracy in impeding private participation, by requiring multiple permits and generally causing delay, whether intentionally or not. This has been a major handicap in Egypt, and deters investment in civilian feeder industries too.

Further obstacles to private participation, both direct and indirect, are the array of official controls that regulate imports of needed raw materials, machinery, and components, and exports of finished products. Once again, these vary from country to country, but form part of the general unease in private-state relations. This is also reflected in an underlying distrust of the state's long-term intentions and the reliability of its attitude towards the private sector; a sudden, arbitrary switch in government policy could nationalise private defence companies or other industrial concerns. That said, nationalisation does not appear as likely a prospect as liberalisation in the Arab arms producing states. But the preceding elements of bureaucratic hindrances, regulatory controls, and funding nonetheless combine to create an inhospitable climate. As one experienced Arab analyst summarised, this is expressed in the fact that the Arab *rentier* economy encourages traders more than industrialists.[10]

Source of Funds

The issue of finance is evidently an important factor affecting participation by the private sector in Arab defence and civilian feeder industries. But in fact it goes well beyond the private sector, to affect military industrialisation as a whole. The most obvious indication of this is the assertion by the Egyptian Minister of State for Defence Production that his country could rely on itself for 60 per cent of its military hardware needs but lacked the necessary funds. Indeed, the impact of financial constraints has been such that Egyptian purchases of equipment and major weapon deliveries dropped from $2.3bn in 1984 to $300m in 1988.[11] The Gulf crisis and subsequent war of 1990–91 may alter this picture, though: the writing off of $7bn of Egyptian military debts owed to the USA and American approval of new combat aircraft sales are likly to defer budget cuts.

Yet the main prospect is of tightening Arab financial constraints, casting doubt on the ability of local governments to fund military industrialisation. Iraq incurred foreign debts of at least $40bn and possibly as high as $100bn during the earlier war with Iran, and now faces even greater reconstruction and reparations bills following its invasion of Kuwait and the second Gulf war. Saudi Arabia has been compelled to adopt deficit spending since the late 1970s, and was even compelled to borrow from Western banks in order to cover its commitments during the Gulf crisis. Further state investment in major industrial undertakings may be impracticable and even unwise for most Arab countries, therefore. This is especially so when the additional investment needed to expand and develop associated civilian industries and technical manpower is taken into account.

At present, state funds account for the greater part of defence industry capital and costs in Egypt, Iraq, and Saudi Arabia. In the first country, the wholly state-owned NODP is responsible for roughly 60 per cent of defence production. Since the state owns at least 51 per cent in the AOI's nine companies – of which only three have a foreign stake approaching the 49 per cent ceiling – this suggests that the state's share probably accounts for another 20 per cent of total investment and recurrent expenditure in the Egyptian defence industry. This leaves 20 per cent of the total that is owned by private capital, part of which is in fact also state-owned and part foreign. Saudi Arabia's ordnance factories are wholly state-owned, and government loans are central to the establishment of private defence companies (though this investment is meant to be recoverable). The Iraqi defence industry of the 1980s, on the other hand, was completely financed and run by the state, as were virtually all feeder industries, in partial contrast to Egypt and Saudi Arabia, where some such facilities are privately-owned.

Most striking is the limited ability of the Arab governments concerned to involve private capital. This is most obvious in the near-total absence of commercial banks or other financial institutions and funds. Indeed, commercial agencies are even reluctant to invest in associated civilian industries. This is due partly to their distrust of the state, based on previous bad experiences, and partly to the general tendency to seek the rapid turnover and quick profits usually attained from other activities, such as construction and commerce. The net result is to limit the volume of funds available for military industrialisation. The low level of private sector participation in turn discourages foreign companies from investing in Arab defence projects.

A further instance of this problem is the inability of Egypt, the leading Arab arms producer, to establish central funds dedicated to the development of the defence industry along the lines of the Turkish Defence Industry Promotion Fund (DIPF). The SIDF fulfils that purpose in Saudi Arabia, but, in contrast to the Turkish agency, lacks a vibrant private or civilian manufacturing sector that is able and willing to take advantage and make its own contribution.

Availability and Skill of Technical Manpower

Qualified manpower is ultimately the cornerstone on which any industrial, or indeed any economic, enterprise rests. The situation in Arab industries varies from one country to another. Egypt occupies the foremost position, in terms of the size of its industrial workforce, number of specialised personnel (engineers, scientists, and technicians), variety, and experience. Iraq comes next, followed by Saudi Arabia. This order broadly parallels population size and sectoral distribution of the economy, though it incorporates certain discrepancies. Thanks to a sustained educational effort to produce more engineers and technicians since the early 1970s, Iraq has built up a sizeable qualified manpower pool. Similarly, despite the weakness of its manufacturing industry, Saudi Arabia has developed an advanced petrochemicals sector and produced a number of highly trained engineers and technicians and a number of associated research and educational institutions.

This summary applies to industry in the Arab countries as a whole; naturally, their defence sectors compete for limited human resources. Given the importance of national security imperatives, however, especially in Egypt and Iraq, the defence sector has been able to obtain much of its manpower needs. Thus the Egyptian and the Iraqi defence industry each employs an estimated 100,000 persons, a sizeable proportion of the national industrial workforce. Saudi Arabia, in contrast, suffers a severe shortage of industrial manpower, particularly in the defence sector, where development plans are specifically designed to produce engineers and technicians (at the Prince Sultan Military City). Both Egypt and Iraq instituted special education and training courses in 1975, following the establishment of the AOI, in which they sent some 2,500 students to the USA and Western Europe. Other academies and universities in all three Arab arms-producing countries have trained additional numbers of engineers, scientists, and technicians.

Impressive as the figures may be, though, they are not a sufficient indicator of the status of Arab skilled manpower in the defence sector. For one thing, a simple calculation of the ratio of manpower-to-output indicates a level of $8,000/worker or a potential peak of $15,000/worker in Egypt (assuming full utilisation of potential, but with no employment increase – an unlikely prospect), compared to a low of $25,000/worker or a more probable $42,000/worker in Israel. (For Israel manpower levels are assumed to range from a peak of 100,000 to a low of 60,000 after the cutbacks introduced since 1985.) However, Israeli output refers only to military products turnover. For Egypt, conversely, the figure for total production of $800m annually includes the 50 per cent allocated to civilian production. There figures also cover all forms of military production, including uniforms, mess kits, and the like. Of course, low productivity has much to do with equipment and organisational bottlenecks (including bureaucratic influence on marketing), but it is also affected by training, manufacturing techniques, and management.

A further point is that Arab defence industry in general suffers from a serious imbalance in the allocation of its personnel, and, by implication, in the pattern of their training. One problem is that there are insufficient numbers of middle-level skilled workers, who are the primary group in acquiring manufacturing technology and transmitting accumulated experience. This problem is a relatively secondary one, though, that has been addressed to some extent within Arab defence industry. The sector suffers a more serious shortcoming:

specifically, while much emphasis is placed on training production engineers and shopfloor technicians, far less attention is paid to the development of managers, designers, and other white-collar personnel. This naturally reduces efficiency and thus productivity, impedes acquisition of technology, and finally raises costs and hinders marketing, eventually reducing the economics of scale for indigenous defence production.

Volume and Importance of Exports

Of the three main Arab arms producers, only Egypt is an exporter. Its export activity is relatively recent, commencing in the early 1980s in response to Iraq's urgent need for combat consumables. According to a variety of unofficial estimates, the total value of Egyptian military exports quickly reached $1bn in 1982, though it has since declined from that peak. Exports were subsequently estimated to run at $500m annually between 1983 and 1987, to be curtailed as Iraq first initiated local defence production and then effected a ceasefire with Iran. By that time Egyptian foreign sales had dropped to $300m at most, and have fallen still further since, possibly as low as $50m.

Although statistics are lacking in both detail and reliability, foreign military sales have accounted for anything between 8.3 and 25 per cent of Egyptian defence production in the 1980s, assuming exports worth $50–100m and a total output of $400–600m annually.[12] (Obviously this allows wide leeway and accepts the most optimistic Egyptian statements.) These rough figures need careful interpretation. The low of 8.3 per cent suggests that exports occupy a minor place in Egyptian defence production; conversely, though substantial, the high of 25 per cent remains relatively modest, and cannot compare to the 50–75 per cent export level (worth $1.5–1.6bn by 1990) achieved by the Israeli defence industry in the 1980s.[13]

What this suggests in the case of Egypt is that there has been a relatively steady and constant level of local demand for defence products, accounting for a major part of output. However, this demand has not expanded, and indeed cannot be guaranteed to continue at its present level given financial constraints and budget-cutting exercises underway at the end of the 1980s. Expansion of production has therefore been linked mainly to external demand, that is, to exports. The rapid growth of the NODP workforce (by 250 per cent in the period 1980–88) and of output generally was directly due to the sudden, massive surge in Iraqi orders.

In other words, while the Egyptian armed forces can be depended on (within existing circumstances and needs) to absorb a certain proportion of production, the local defence industry cannot utilise a greater part of available capacity without increasing exports. And with capacity estimated to be worth $1.5bn annually, this consideration is a major one. Further confirming the linkage, it was the re-exporting of army ordnance stocks and guaranteed sales of such items as the *Tucano* aircraft that provided the Egyptian defence industry with the funds for new expansion and projects. Conversely, as exports have dropped at the end of the 1980s, so has total output.

A further serious qualification is that much of the ordnance shipped by Egypt to Iraq, especially in the first years of the relationship, came out of the Egyptian Army inventory and so did not reflect productive capacity. Thus SIPRI estimated that the domestically-produced share of total Egyptian military exports (excluding what were effectively re-exports) was no more than 5 per cent in 1981–85.[14] The assertion that the NODP workforce witnessed a 250 per cent expansion in 1980–88 does imply, though, that the Egyptian industry quickly geared up to meet demand and exploit the opportunity for making profits and expanding itself.[15] The AOI was moreover enabled to launch its *Tucano* trainer programme, assured of the sale of 80 aircraft to Iraq.

So Egyptian export performance has had a mixed record. On the one hand the sector has brought sorely-needed funds, which have largely been reinvested in the defence industry, while on the other it faces severe difficulty in sustaining itself. One reason is that Egyptian products are often not much cheaper than alternatives on the world market. In fact, the UAE cancelled a $3.5m order for *Fahd* APCs and *Amun* AD systems in late 1988 because they were deemed to be over-priced.[16] The *Tucano* trainer aircraft sold to Iraq cost as much as the originals supplied by Brazil, while the American administration expects Egypt to produce M-1 tanks at a final price that is $1m above the purchase price direct from the USA.

Political and technical factors have also intervened to prevent further sales of major weapon systems to other Arab states; examples are the preference displayed at various times in 1986–89 by Saudi Arabia, Oman, and the UAE for *Hawk* and *Tornado* aircraft offered by their European manufacturers, instead of the *Alpha Jet* and *Mirage* 2000 versions made in Egypt under licence. The net result has been to shorten the *Alpha Jet* production run and to prevent the hoped-for transition from parts manufacture to assembly in the case of the *Mirage* 2000. This pushed up the unit cost to Egypt, undercutting locally-added value and so undermining economics of scale as well as reducing technology transfer. Finally, Egypt suffers additional problems in marketing, and even more so in offering financial incentives and offsets, both of which have become essential inducements to attract Third World arms customers.

These drawbacks of poor export performance have a significance that goes beyond Egypt to affect Iraq and Saudi Arabia. The prospects for Iraq of restarting the local arms industry or exporting its products following the second Gulf war are poor indeed, but before the events of 1990–91 exports promised to form an important additional source of revenue that could have allowed further expansion of the sector. The lesson applies especially to Saudi Arabia, where the state is encouraging the emergence of a private defence sector. There private companies rely absolutely on sales – of which exports form a major component – unless they are to be heavily subsidised, which would be a self-defeating exercise. In general, any attempt in Saudi Arabia or Egypt (or Iraq, hypothetically) to involve the private sector will therefore depend on increasing the volume of exports as a measure of success and viability.

Conclusion

Arab military industrialisation is taking place in an international context that is marked by the proliferation of industrialised and Third World arms producers. Intensified commercial competition globally is, moreover, accompanied by the accelerating pace of technological change. These trends confront emerging Arab arms makers with ever greater challenges, which place increasing pressure on the components and linkages of the military industrial system as a whole. Survival demands advanced military R&D (possibly including greater technology transfer by foreign investors eager to improve sales by subsidiaries/partners), organisational development (managerial and marketing skills), and industrial rationalisation and restructuring (diversification into support hardware and services, and into civilian production).

A number of criteria will determine how successful Arab arms industries will be in maintaining their viability and in making the transition from current capabilities, therefore. Among the determinants to be discussed in subsequent chapters are organisational structure, manpower development, technology transfer, availability of funds, and the parallel growth of the civilian scientific and the industrial base.

9

Assessing the Economic Impact

Security and other strategic or political incentives may have been the pre-eminent factors promoting Arab military industrialisation to date, but the impact of local defence production on national economies still needs to be examined. Put differently, how relevant or valid have the classic economic arguments been in favour of military industrialisation? After all, Arab officials have repeatedly cited economic incentives and financial benefits in support of establishing indigenous arms industries, which seems to imply that economic thinking does underlie their creation.

Indeed, the economic dimension is important, if only to the extent that it constrains military industrialisation. The main economic consideration for decision-makers is not whether or not local defence production really aids development, but whether or not it can attain enough commercial feasibility to make the setting up of a local arms industry supportable. In stark terms, the issue is simply one of finance. Thus the core economic issues effectively drop out of the debate, and the question becomes one of financial ability to implement programmes determined primarily by perceptions of national security. Whether funds are then found thanks to the availability of surplus capital or through commerce makes little difference at this level.

So what discernible impact has Arab military industrialisation had on local economies and on development, in terms of the classic economic arguments made in support of this sector?

Effect on Employment and Manpower

In global (pan-Arab) terms, the defence industry employs some 200,000 persons (of all levels and skills) compared with an estimated grand total of 15 million persons in the Arab industrial workforce.[1] This may seem an insignificant proportion, but it gains in importance when contrasted with the industrial workforces of Egypt, Iraq, and

Saudi Arabia individually. Statistics are lacking in the first case, but figures for 1986 indicate 193,392 and 130,487 workers in the Iraqi and the Saudi manufacturing industry, respectively.[2] An arms industry of virtually any size – it is reported to be 100,000-strong in Iraq – will therefore employ a substantial part of the local industrial workforce.

The foregoing suggests that, while local defence industries are likely to have little impact on total employment levels, they may compete for industrial labour. So the pertinent questions are the following: do Arab arms industries provide employment for categories of skilled labour which might otherwise lack jobs, possibly because insufficient investment is being made in civilian industries? Or, conversely, do they in fact divert trained technical manpower from other sectors of the civilian industry and national economy? The answers appear to point both ways.

On the one hand, there is a surplus of technical and professional manpower in the Arab World generally. Indeed, the lack of job opportunities has, for many years, combined with other social and political factors to propel a constant brain drain towards the West. Unemployment among college and university graduates is a major phenomenon in many Arab countries. Thus any productive enterprise should in principle be economically beneficial, especially if it encourages expansion in associated manufacturing activities (steel and chemicals). Additionally, industrial projects launched to cater for special defence needs (electronics and computers) help to employ and retain within the country certain categories of scientists and experts who might otherwise change career direction or emigrate.

The defence industry also employs former officers and military engineers with accumulated experience that might otherwise be lost. (Incidentally, an implication of the existence of surplus labour is that the defence industry is not quite as compelled to raise productivity or capital-intensive techniques for economic and commercial reasons, and so can afford to be relatively labour-intensive.)[3]

On the other hand, the defence industries do compete with the civilian sector for human resources to some extent, especially for certain personnel categories (such as managers, designers, and specific skilled manufacturing workers). The industry is by nature a relatively technology-intensive one, and accordingly requires a higher proportion of the same qualified manpower that is also needed in the civilian sector. It does not help matters that a sizeable section of the Arab technical manpower pool cited in official statistics – particularly professionals (engineers and white-collar personnel, such as administrators) – is in fact poorly-trained or under-qualified for industrial activity. Thus competition for qualified manpower becomes stronger, most of all with the civilian *manufacturing* industry. The defence industry might find that it can afford to be labour-intensive at certain levels and in particular branches of its activity, but it must achieve economies of production through improved organisational and operational skills – management, in-house training, product/system/production design, computerisation, research, and marketing – which require specific types of qualified person.

However, in the absence of more detailed government industrial statistics and studies, and given the addition to the manpower pool of tens and even hundreds of thousands of Arab technical school and university graduates in the past two decades (owing to the enormous effort invested in education and training), the conclusion here is that the diversion of technical manpower into defence industries does not in itself form a major impediment to national economic development. Indeed, given the investment and appropriate training schemes – that might also attract Arab expatriate professionals to return – the defence sector could even contribute favourably to the pool of Arab trained manpower.

Stimulation of Associated Civilian Industries

Left over from the preceding section is the question of whether the defence industry creates new jobs and trains additional manpower in the civilian sector. This is part of the broader issue of the role of military industrialisation in stimulating associated civilian manufacturing activities. That is, the defence industry supposedly creates new jobs and undertakes training within itself, but does it indirectly do the same for other sectors?

As in the case of employment, the impact of military industrialisation on the civilian sector may in principle be beneficial. More to the point, there is no evidence that the growth of Arab defence industries has had a direct detrimental effect on civilian industry (the issue of diversion of financial resources – (state funds in particular – from development will be discussed later). However, the matter is not one of searching for possible negative effects, but of assessing the validity of the standard arguments in favour of indigenous defence production on grounds of its economic benefits. Thus it is virtually inevitable that the growth of Arab defence industries has had the positive result of increasing demand for local production of non-fuel primaries and raw or worked materials, unfinished goods, and parts and components. But has this led to major growth in the civilian sector, and has it been able to supply an appreciable part of that demand?

Here the broad impression is that military–civilian industrial interaction has been limited, owing in large measure to substantial technical and structural problems in the associated civilian sector.

Some of the products required for defence production, including primaries, are specialised and need dedicated manufacturing facilities (and equally specialised personnel). This requires additional investment, but, as the market is necessarily limited, under-utilisation of capacity and high production costs result.

This might not be the case if supply and demand were rationalised through inter-Arab co-operation; but for individual Arab arms-producing states the problems are considerable. Many of the products required by the defence industries do not fall within this specialised category, though, and conceivably could be supplied economically. However, demand in civilian markets already outstrips local production capacity (regardless of how efficiently that capacity is used or its products marketed): further expansion is dependent on more investment and training, and on solving the basic obstacles impeding Arab industrialisation generally.

Among the obstacles are the problems of bureaucracy and poor planning in state industries, and the range of disincentives facing the private sector. The last particularly, but not solely, is affected by administrative red tape and government restrictions, and by custom duties and import barriers which may even make local products more expensive than equivalent imports. (It must be remembered that all these comments apply especially to the very countries most involved in defence production: Egypt and Iraq.) Indeed, the lack of incentives for private participation in associated civilian industries affects the defence sector, especially in high-tech areas such as electronics and computer technology, which are so central to both manufacturing processes and military end-products. And, as mentioned previously, the reluctance of private capital to invest in major industrial projects (in conjunction with the state or independently of it) deprives the defence industry of major funds. Thus Arab defence production still involves a relatively high degree of imports (of all types, but increasingly as the manufacturing and technological component of the items rises).

Military industrialisation appears to have had, at best, a modestly stimulating effect on associated Arab civilian industries. The implication is, though, that this has been due to structural problems which have prevented the latter from responding to demand, rather than to inherent limitations to the emergence of a dynamic interaction. That is, the defence sector could have a more stimulating effect on civilian industry if the latter were free to respond and grow. By the same logic, the ability of Arab defence industries to stimulate job creation or technical training is dependent on the development of the civilian sector. A further implication is that although a simple strategy of import substitution might not be the answer for the associated civilian (feeder) industries, internal reform and improved planning would allow a healthier mix of imports and local products to be supplied to the defence sector.

Military R&D Spin-Offs for the Civilian Sector

Subsumed under the role of military industrialisation in stimulating civilian industries is the question of how the latter benefit from defence-related R&D and technology acquisition. According to this view, products, materials, and manufacturing processes developed at specialised centres can be used in both military and civilian industrial activities. Again, this appears to be a sound proposition, but it is among the least applicable in the Arab case. Most importantly, Arab military R&D effort is minimal, and the number of centres engaged in technological studies is extremely small. Indeed certain areas of research are not even touched upon at all, owing to lack of funds, personnel, and relevant technology.

Furthermore, there is no evidence of civilian industrial products being developed specifically as a result of military R&D in the Arab case. In fairness, the connection is limited even in other arms-producing countries (including the AICs), as many of the products and materials under development have little utility in civilian goods – tritium and hardened steel or frequency-hopping radios and decoders are examples. Thus much of the interaction, in fact, involves the adapting of production lines to enable factories to produce both civilian and military goods of similar quality. The issue here becomes one of using certain areas of R&D to improve manufacturing processes and quality control – itself a useful and even vital function. However, it seems that in the case of the one arms-producing Arab country that uses part of its military industrial capacity to produce civilian goods – Egypt – there is little actual interaction as such, and the effort reduces essentially to using available machinery that would otherwise lie idle.

Finally, the broad starting question needs to be readdressed: does the defence sector act as a base for advanced industrialisation in the Arab countries? Evidently this refers not to production within the defence industry, but to the latter's role in assisting the expansion of the civilian sector's industrial and technological base. The assumption being that, through local R&D or imports, the defence sector will introduce modern technology (products and systems) and manufacturing processes into the country, as well as creating demand for local feeder industries of commensurate quality, and then eventually starting a snowball effect.

As explained earlier, though, the civilian basis for such a dynamic process is non-existent or weak in the Arab case, and so is impeded in benefiting from the growing demand for R&D spin-offs of the defence industry. What stimulus there has been for greater industrialisation and technological advance in the civilian sector has instead mainly come from heavy demand in civilian R&D in its own right, and of seeking those areas of military R&D which might also be beneficial to civilian industry. But at present, the modesty of Arab military R&D effort and under-development of civilian industries precludes any effective technological interaction.

Foreign Currency Savings and Reduced Imports

The argument that foreign currency savings can be made and imports reduced – leading to an improved balance of payments – is possibly one of the most attractive, yet most deceptive, economic tenets in favour of indigenous military industrialisation. The central thesis is, of course, that local manufacture supplants the need to purchase weapon systems from abroad, and so reduces the need for hard currency.

At one level, this should hold true for the defence industries of Egypt, Iraq, and Saudi Arabia, which provide their countries with the bulk of their ammunition needs, as well as a descending proportion of small and medium weapons and supporting equipment, at least until the second Gulf war. Egypt and Iraq moreover, benefited in the 1980s from tank modernisation programmes that reduced the need for new, imported vehicles, and also manufactured a number of gun and rocket artillery systems. Egypt (which would have been followed by Iraq had it not been for the second Gulf war) assembled a number of aircraft locally, or produced parts for them, while all three Arab producers expanded into electronics and communications equipment. To this extent, Arab defence industries were supplanting imported goods, wholly or partially, and so presumably were cutting down on hard currency spending.

The argument requires closer scrutiny, however. On the one hand, it holds true for as long as the final product can be made or sold at a total cost not exceeding that of the equivalent import. Of course, it should be understood that the indigenous defence industry still has to import many component parts, for which it pays in hard currency, in order to produce any end-product. The sum total for components (including materials) should logically be less than the price of a complete product imported whole, if any portion of assembly or manufacture is done locally, since the cost of local labour is low. (It is the labour that allows the greatest economy in local defence production, rather than any other element, such as local production of raw materials and primaries.)

This appears to have been the case with the Egyptian copy of the *Tucano* trainer aircraft, 80 of which were sold to Iraq at exactly the same price as the Brazilian export version: local costs in Egypt must have been at most equal to, if not less than, production costs in Brazil. Thus Egypt can be assumed to have saved in foreign currency an amount equal to its locally-added value (principally labour, plus any locally-manufactured parts involving further labour or cheaper materials); and indeed Egyptian officials (and foreign investors) point especially to low labour costs as the country's principal advantage.

However, this argument runs into certain difficulties, not the least of which is the need to account for the initial capital invested in setting up the original aerospace industry (or specific factory) and training its personnel. That is, the final cost of producing specific weapon systems or non-combat equipment should be measured in terms of total investment, a large part of which involves importing machinery, manufacturing tools, designs, and construction/production expertise. In fact, the need to construct or adapt new production lines, provide the necessary tooling, and train local personnel may raise the cost of local assembly or parts manufacture to the extent that the sum total of the parts (including imported components) does exceed the original price of a system imported whole. Furthermore, technology transfer and licences also cost money, which has to be paid to foreign suppliers and thus added on to the cost. Similarly, profits may have to be shared with licensers, and there often is an additional cash outflow if industrial projects or joint ventures are set up under special financial and commercial arrangements – tax exemptions, low tariffs and customs duties, or in free-trade zones.

The counter-argument to the economic benefits achieved through foreign currency savings and reduced imports is made most obvious by Egypt's proposed M-1 tank programme. Accepting American administration figures at face value, it will cost Egypt $4m to produce each M-1 locally – in a production run of 550 – compared with the $3m price-tag on a direct import from the USA (or the $2.8m cost per tank to the US Army). So, even assuming that Egypt succeeds in raising its own contribution to a maximum of 19 per cent by value of total production cost (550 vehicles at $4m each), the import share will still be $110m greater than the cost of buying the same number of ready-made tanks from the USA at $3m each. Evidently, some savings could be made by extending production runs if there were guaranteed export orders, but this is not in the offing. Moreover, part of the Egyptian share of production will, in fact, be undertaken in association with American companies, so the final local contribution will be less than stated. The implications are obvious: hard currency will not be saved, nor will imports (of components and machinery) and thus the total cost be reduced.

In assessing this economic argument, therefore, a crucial element is to take into account the fact that nascent defence industries must invest heavily in constructing factories and production lines, and even then must buy additional tooling and machinery whenever new products are planned – all of which must be imported, using hard currency. In the contemporary Arab context, this means that military industrialisation was a particularly costly effort for Iraq, which set up a major arms industry in the mid-1980s at a rapid pace. Whereas Egypt built up much of its current infrastructure in stages from the early 1950s onwards. And even once the initial investment has been made, the import content of local defence products will remain high. Indeed, given heavy reliance on imports of raw materials as well, it is more than likely that even such basic items as ammunition can be purchased abroad more cheaply than can be produced locally. After all, despite possessing a local capability for many years, Egypt was not moved to increase the manufacture of ordnance and substitute imports until massive Iraqi demand allowed expansion of production facilities and workforce at the NODP in the 1980s.

The central issue, in conclusion, is to calculate costs and benefits for the Arab defence industry on a national, rather than a factory or project basis. Because most Arab military factories are state-owned, however, publicised current costs tend to be calculated on a project basis, ignoring the original or total investment. Of course, if the defence product were being exported, all costs (including licence fees, for example) would have to be factored in; but when making statements officials might refer to project cost only and so portray the item as being cheaper than it really is. Yet since the economic argument purports to show the benefits to the country as a whole of indigenous defence production, it needs to be assessed at that level too.

The first concluding point to be made is that in an age of fiercely competitive world markets and the technological revolution, any defence industry will be highly reliant on imports of hardware and know-how, and so classical notions of import substitution are obsolete and impractical. That need not necessarily lead to the opposite situation, though, in which the local defence industry is built up partly as a subsidiary of foreign companies or in effective free-trade zones (FTZ); in such a case the whole issue of hard currency and imports loses meaning, owing to the extensive openness and interdependence between the national economy and local industrial effort on one hand, and foreign investors and the world economy on the other.

The preceding scenario represents to some degree what has occurred in the east Asian newly-industrialising countries (NICs) with their dynamic defence industries. But their example leads to the final crucial point: in order to attain a balanced position between these two options while achieving some hard currency savings and import reductions, Arab arms producers need to raise local content through increased R&D effort and expanded civilian feeder industries. This dynamic has been at the heart of the east Asian success story. The R&D effort will at least increase the number of products and components that can be manufactured locally (and, it may be hoped, with less reliance on imported designs and know-how), while a dynamic civilian sector will provide many of needed materials and parts at reduced cost by increasing the share of local labour.

Export Growth

The overall scale of Arab military exports, and thus their impact on trade balances, has been at best modest. In the case of the one real Arab arms exporter, Egypt, the total value of exports of all kinds has been in the range of $4–5bn annually since 1982, during which time the share of manufactured goods (of all types) has risen steadily from $490m to an estimated $1.16bn in 1988.[4] Assuming $50m worth of defence exports annually, military sales to foreign customers appear to have accounted for 1–1.25 per cent of all Egyptian exports, or 5–10 per cent of exports of manufactured goods. Given the lack of accurate data, these percentages offer little more than a rough guide, but they nonetheless serve to show the modest ranking of defence products among exportable items in Egypt.

Nor are prospects any better for Iraq and Saudi Arabia. The former will take several years before it overcomes the effects of the devastation wreaked on it during the Gulf war of 1991, and is unlikely to emerge as an arms exporter even if it rebuilds its local defence industry. With exports worth $24.3bn and manufactured exports worth $3.74bn in 1988, Saudi Arabia is as unlikely to develop an arms-exporting capability, even assuming it could reach the technical levels needed.

Increase in GDP

Linked to both imports and exports, but distinct from them, is the issue of the contribution of defence production to GDP through locally-added value. Essentially, value is added locally through the use of indigenous labour in the manufacture and assembly of military systems; additional contribution comes from the labour and capital investment embodied in such components and parts as are also manufactured locally and from the utilisation of raw materials extracted and processed within the country.

By the same token, the degree to which value is added locally also indicates the level of utilisation of national natural and human resources. Thus, whether the country still imports much of its combat *matériel* or achieves significant defence exports, that proportion of its indigenous defence production which is actually supplied to the armed forces represents a net contribution to GDP and so supposedly enhances the national economy. (Naturally, any exports would form an additional contribution.) Put differently, as long as a given government deems it necessary to spend on defence, it can contribute to GDP by using part of national military expenditure to purchase certain types of hardware and services from local defence industries.

For a variety of reasons, this aspect is one of the most difficult to quantify and thus assess for the Arab countries. Real defence expenditure is difficult, and often impossible, to estimate for most of them with any confidence. The same applies to the breakdown by category (between acquisitions, pay, maintenance, and so on). This is significant because certain expenses may be covered from outside official defence and state budgets through loans, grants, and extra-budgetary support (such as enjoyed by Egypt in various phases, or Iraq through borrowing in 1980–88). Without accurate information the share of defence (and defence production) in GDP cannot be calculated properly, since some items may be funded totally extraneously, leading among other things to an artificial picture of the defence burden.

That said, some tentative observations may be made about the Arab arms producers. According to official claims, actual Egyptian defence production is worth some $400m per annum, and may have reached as much as $600m in 1985 (depending on how the figures are understood). The defence industry moreover produces civilian goods, supposedly worth an additional $350–400m annually. Total production in the sector is thus valued broadly at some $800m, although its industrial capacity is set at $1.5bn.

While stressing once again that these figures must be viewed with great caution, they nonetheless point to a considerable contribution to GDP. The latter was estimated at $29.58bn at factor cost in 1988 (or $31bn including net indirect taxes), indicating that the defence industry's total share (both military and civilian products) stood at around 2.67 per cent (assuming production unchanged at $800m). More importantly, the industry accounted for 9 per cent of total industrial output and an estimated 14 per cent of total manufacturing output.[5] Thus although there is no satisfactory way of ascertaining whether local defence production is cost-effective in economic and commercial terms, its contribution to Egyptian GDP is substantial by normal standards, based on stated output.

Complete absence of comparable statistics makes it impossible to evaluate the contribution to local GDP of the defence industries in Saudi Arabia and Iraq up to 1990. The most that can be safely ventured is that, with GDPs in 1988 of $75.39bn and $46bn (estimated), respectively, the contribution of local defence production would have been modest for the foreseeable future (even leaving aside the virtual destruction of the Iraqi economy during the second Gulf war) although the cost of developing most programmes (especially nuclear) was prohibitively costly.[6] The same may be suggested on the basis of total industrial output, which in 1988 stood at $32.22bn for Saudi Arabia (no reliable estimates exist for Iraq). However, the preceding two items were exaggeratedly large owing to the oil sector, and when the comparison is made with the far more modest *manufacturing* output – $6.17bn of value added in Saudi Arabia and $5.4bn in Iraq in 1988 – it becomes plausible that defence production could have occupied a prominent place within local manufacturing industries.[7]

The twin examples of Saudi Arabia and Iraq in the 1980s are particularly useful in highlighting the special issues connected with Arab military industrialisation in this context. They will both remain oil-based economies for the foreseeable future, and rely on oil exports to fuel development in other sectors, including industry. Thus the oil sector will continue to play a predominant role in forming GDP for a long time to come. For any other economic sector to assume a major (albeit not the primary) role, it must experience massive, sustained expansion in the years to come.

In the case of the defence industry, therefore, it can only make advances as part of the expansion of the entire industrial and manufacturing sectors. This linkage has already been repeatedly suggested in other connections, and here gains an additional argument in support. It is further reinforced by the fact that the main way for Arab defence industries to increase their contribution to GDP is to add to their locally-added value through development of civilian feeder production (and more generally through 'deeper' production, thanks to R&D, and greater interaction with the civilian sector). In the Saudi and (potentially) the Iraqi case, moreover, this process can be assisted by developing special branches of the civilian industry, such as petrochemical products, that can be used in defence production, though the general lessons are equally relevant to Egypt.

Reduced Defence Burden on State Budgets

In providing military products for the national armed forces, local defence industries are assumed to reduce the financial burden of defence on state budgets (and implicitly on state-funded development). This is supposedly achieved by cutting down the price that would have had to be paid for equivalent military imports, and also by providing additional revenue through defence exports or civilian production. Generally speaking, the greater the share of locally-added value in the local military industry, the bigger the reduction in the defence burden.

Assessing the validity of this argument in the Arab context is yet another thorny issue, owing to the total absence or unreliability of official data. As described earlier, Egyptian figures do not include acquisitions, nor do they account clearly for foreign assistance. Iraqi statistics are completely unavailable, while Saudi Arabia publishes budgets for recurrent and non-hardware items only. With the inadequacy of figures on defence production, it becomes wholly unrealistic to guess at its impact on state expenditure.

What can be hazarded, nonetheless, are a number of basic propositions about the impact of defence production on defence expenditure generally. The first of these, as former Egyptian Minister of Defence and Defence Production Amin Huwcidi notes, is that the impact is connected with the level of arms imports in a given country, which is in turn connected to its national security situation.[8] Thus in the case of the Arab 'confrontation' states and Israel, where defence expenditure is high, local military industry faces a far greater challenge in order to supply a significant share of needs; but for the same reason can also have a major impact on defence expenditure if it succeeds. At the end of the 1980s, Israeli defence factories were providing the armed forces with products worth \$500–600m annually, out of total sales worth \$2–2.2bn.[9] Local demand therefore represented some 10 per cent of the total defence budget, and at least 15 per cent of total acquisitions. (These figures do not include non-military goods produced by the defence industry for local and foreign markets.)

Conclusion

The picture presented here of the contrast between national security and economic incentives may change in the future. The economics of defence production might become more important as the urgency of the national security imperative recedes. Furthermore, the second Gulf war has stimulated enormous interest among developing countries in high-tech armaments: this will demand an even higher level of technological sophistication from Arab defence industries, and so to attain a modicum of commercial feasibility will become increasingly necessary for their viability and survival. They must improve their competitiveness in regional and world markets, or incur an ever-growing cost in the effort to keep pace. Just how successful Arab defence industries will be depends on a number of factors, such as the qualifications of the local workforce, Arab investment in R&D, and the role of the private sector.

The implications are most applicable to Egypt and Iraq, with their established industries, national security policies, and large defence establishments. In contrast, Saudi Arabia may simply continue to postpone large-scale military industrialisation – especially if its post-Gulf war assessment is that it no longer faces any imminent external threat – unless clear commercial advantage is to be had (which is not altogether likely). A further, core implication here is that, in view of the general underdevelopment of the Arab scientific and industrial base and structural imbalances in national (domestic) economies, the process of enhancing commerciality and so improving the economics *of* defence production will eventually mean crossing the dividing line into seeking to implement the economic arguments *for* defence production.

Judging from present trends, it would appear in conclusion that for the foreseeable future economic considerations will not be decisive, and that military industrialisation will not be measured substantially by its contribution to development in the countries concerned. Indeed, given the current underdevelopment of their civilian economies (including overall structure, industry, infrastructure, and the sciences), investment in defence production can hardly be predicated on progress in both the military and the civilian sector. Thus, instead of justifying military industrialisation on the grounds that it can spur growth in the civilian economy, the real question is how far Arab defence industries can progress without significant improvement in the relevant sectors of the civilian economy.

But once again, in the light of the high priority given by Arab governments to national security and defence, this outlook is unlikely to reduce investment in military industrialisation (which is disproportionately high compared with other sectors) under normal circumstances. In fact, continued concentration of state resources (financial and human) in the defence industries may allow them to develop at a faster rate than their civilian counterparts, despite the handicaps, and the two sectors will tend, as at present, to be largely dissociated in their growth.

10

Arab Scientific and Industrial Base

To complete the assessment of Arab defence industries, this chapter reviews in greater detail three vital prerequisites of effective military industrialisation: the civilian industrial base, the R&D base, and scientific and technical manpower.

Industrial Base

The importance of the national industrial base to any country's defence sector stems principally from its general role in training technical manpower, creating a conducive climate and supportive 'culture', and providing feeder goods, but also from the contribution of specialised branches. Heading the list among the last are the automotive, electrical machinery, non-electric machinery, and electronics industries. Also of primary importance, as feeders, are the iron, steel, non-ferrous metals, and metal components sectors. Indeed, these industrial categories are taken to represent the major indicators – along with the share of manufactures in GDP and the ratio of scientists and technicians involved in R&D – of the state of a given industry and its ability to support a dynamic and viable defence sector.[1]

No attempt is made here to provide an exhaustive and authoritative review of the Arab region's civilian industrial effort. Rather the aim is to offer a few indicators of the nature and scale of manufacturing industry generally and of particular branches. According to the *Unified Arab Economic Report* of 1987, manufacturing industry accounted for 11 per cent of total Arab GDP in 1986, or $42.4bn at current prices, of which 89 per cent ($37.9bn) was generated in the oil-rich countries.[2] (As in the case of defence spending, available statistics on Arab industrial output and manpower are extremely erratic, and the only year for which reasonably comparable figures are available is 1986.) Table 10.1 shows a detailed breakdown by country, but most relevant here is to note that in the three principal arms-producing states the manufacturing industry-to-GDP ratio is: Egypt, 15 per cent (worth $8.12bn); Iraq, 10.3 per cent ($4.72bn); and Saudi Arabia, 9.3 per cent ($7.08bn).[3] Part of this output is exported, at a ratio (manufactured exports to total manufacture) in 1986 of: Egypt, 16.44 per cent ($790m): Iraq (not available): Saudi Arabia, 26 per cent ($1.82bn).[4] (Presumably these figures include defence industry output, though compilers probably face immense difficulty in calculating it.)

TABLE 10.1 Manufacturing industry in Arab countries, 1986 ($m, current prices)

	Production	GDP (market price)	Production as percentage of GDP
Algeria	7,625.3	60,622.5	12.58
Bahrain[a]	574.5	4,739.4	12.12
Djibouti	61.9	755.0	8.20
Egypt	8,120.0	53,843.9	15.08
Iraq	4,719.2	45,718.6	10.32
Jordan	554.2	4,612.2	12.02
Kuwait	1,295.0	17,657.1	7.33
Lebanon[a]	47.8	378.2	12.64
Libya[a]	1,274.4	20,059.1	6.35
Mauretania	44.2	833.7	5.30
Morocco	2,810.4	15,113.3	18.60
Oman	382.3	7,664.6	4.99
Qatar	414.0	4,958.1	8.35
Saudi Arabia	7,075.9	75,825.3	9.33
Somalia	85.1	1,593.1	5.34
Sudan[a]	572.8	5,853.2	9.79
Syria	1,573.0	21,637.2	7.27
Tunisia	1,216.6	9,420.3	12.92
UAE	2,692.1	20,899.6	12.88
Yemen, North	237.6	2,674.7	8.88
Yemen, South	95.3	1,010.4	9.43

Source: Unified Arab Economic Report 1987, Appendix 1/2.

Note:
[a] At factor cost.

Unfortunately, detailed breakdowns by branch of manufacture are not available for many Arab countries, or are incomplete, making it impossible to draw a regional picture or even a reliable local one for individual states. Nonetheless an impression can be drawn of the scale and nature of certain activities, through random examples. A case in point is the automotive industry, which is considered a primary basis for the emergence of military automotive ventures. In 1988, total production of civilian cars in Algeria, Iraq, Egypt, Morocco, Saudi Arabia, and Tunis reached 83,000 units per annum.[5] Expansion plans underway the end of 1980s were supposed to raise this level to 103,000 annually, though production remained at one-third to one-half of capacity. In Egypt, sales in 1991 of cars assembled by the AOI were down to one-third of its 35,000-car capacity p.a., despite attempts to diversify models and types.[6] Iraq later confirmed additional plans to produce trucks, vans, buses, and 90,000 cars under licence from Mercedes-Benz and General Motors, as well as some 15,000 agricultural tractors of Czech design.[7]

Noticeably, most Arab automotive production to date has consisted of assembly, with varying degrees of local manufacture of parts and components.[8] However, the Iraqi plans of 1989 indicated the growing trend towards complete local manufacture, which was to include engines. This approach was echoed in Egypt, where the AOI hoped to produce a car, in which local input would reach 75 per cent in three years and 100 per cent in five.[9] However, the fact that there are over 12 million vehicles of all types in operation in Arab countries indicates how far the industry falls short of meeting a significant part of local demand.

A broadly similar situation exists with respect to electrical products, such as radios and television sets, production of which is undertaken in a slightly wider group of Arab countries. Here again the focus is on assembly, using mainly imported components, and demand far exceeds local supply capacity. Other, more advanced electronics products are scarce, and tend to be produced at defence-related plants. The computer sector is especially underdeveloped, with near-total reliance on imports.

A final example of the scale and scope of Arab manufacturing industries is the iron and steel sector. Here progress has been made, and much more is envisaged after the implementation of new plans, though the total volume of production is inadequate. Among the more ambitious plans at the end of the 1980s were those of Iraq, which hoped to increase iron and steel production from its level of 1.2 million tons annually to 5 million during 1991–96.[10] However, little of this was of the special steels required by defence industries, and all Iraqi development is in complete jeopardy following the second Gulf war. Egypt is planning to produce special steel at a joint venture with South Korean capital at a rate of 110,000 tons yearly; in 1988 the country produced only 8,000 out of 58,000 tons consumed locally.[11] For its part, at least one plant in Saudi Arabia was producing some 36,000 tons of thermal carbon black annually in the mid-1980s.[12] Two other Saudi plants produced a total of 980,000 tons p.a. of construction bars for use in reinforced concrete.[13]

Interestingly, in its feasibility study for the proposed Egyptian-Korean steel venture, an Austrian design bureau suggested that the ideal arrangement would be to base the project in several Arab countries, and named Algeria, Iraq, and Saudi Arabia to house gradual expansion. This indicates awareness that the natural method of establishing a large-scale industry, and the most feasible economically, would be to do so on an inter-Arab basis. This would rationalise production, pool resources, and both expand and better serve local markets. It would also meet the rule of thumb that the per unit cost is far lower at the large plants that could be set up to cater for a unified market, than at the smaller plants designed to meet local national needs only – there could be a 40 per cent drop in price between a 1-million-ton capacity steel plant and a 50,000-ton one.[14]

Yet in fact, a review of all Arab-Arab or even Arab-foreign joint ventures involved in manufacturing and registered in the 1984 directory compiled by the Arab League and OAPEC shows a mere 17 companies. Of these, five involved iron and steel and five involved aluminium; but several were, in fact, small concerns with limited capacity, or directed mainly to providing rolled products or frames and bars for the construction sector.[15]

One of the negative consequences of the lack of inter-Arab industrial co-operation has been to make individual country-based industries less economical to establish and run. There are additional disincentives, however, which have reinforced this negative pattern at both the regional and national levels. Birks and Sinclair identify a number of problems in their discussion of the Egyptian economy:

> Production [in the manufacturing sector] has been hampered by under-utilisation of capacity with inefficient employment and pricing policies controlled by the Ministries until 1975 . . . Some recent attempts have been made to decentralise decision-making, but not on key issues such as investment strategy, labour force and pricing policy. Most industry is import-substituting, and relatively small in scale . . . Egypt's industrial sector remains unfortunately structured. It caters for the domestic market rather than exports and so is constrained quickly by foreign exchange shortages. Raw materials come from abroad, with the result that shortages of foreign exchange often check expansion.[16]

The *Unified Arab Economic Report* of 1987 makes additional comments, connecting the lack of inter-Arab co-operation with specific obstacles to industrial development within individual countries. It identifies the main impediment as:

> . . . non-implementation of the principles of economic co-operation

among the Arab countries, leading to the continued weakness of productive efficiency in this sector and to its inability to develop itself, due to the denial of the element of large-scale production. Also among the obstacles is the lack of clarity of each industrial strategy in the Arab states, which leads to each industry being assessed from a narrow national [country] viewpoint with no regard to the possibilities of industrialisation for export within the framework of a complementary Arab plan.[17]

It then goes on to note that, even when feasibility has been evaluated in designing individual national industries, their development has suffered from the inadequacy of transport and the existence of customs controls that bar inter-Arab trade. The *Unified Arab Economic Report* of 1989 adds that the policy of hiring expatriate labour also raises costs of local manufacture, as docs the massive under-utilisation of capacity. In the case of iron and steel production, for instance, the use of capacity varies from a low of 17 per cent to a high of 60 per cent for intermediate products, depending on the factory or country, and runs at an average of 66 per cent for all Arab countries in the case of finished products.[18] As endemic a problem, moreover, is that the Arab *rentier* economy ensures higher, secure income for merchants and middlemen than for industrialists, thus effectively discouraging the latter.[19]

Scientific and R&D Base

Although several main elements are necessary for the implementation of industrial development, and, indeed, are sufficient to initiate industrial activity, R&D represents the key to lasting sucess. This is especially so if any given country aspires to achieve a modicum of self-sufficiency or competitiveness and to be more than a mere adjunct to the multi- and trans-national companies of the modern world. The scientific base – that is, science (as knowledge), scientific resources (material and human), and official and public attitudes to science – are a major determinant of the levels of industrial manpower, productivity, diversity, sophistication, and profitability. The existence of an indigenous R&D effort is moreover a precondition for the absorption of foreign technology and manufacturing know-how acquired through local licensed production. By the same token, it is also an absolute requirement for any transition to the production of high-tech products, which are in turn increasingly necessary both as components in larger systems and as exportable commodities in highly competitive world markets.

Given this definition, if a single Achilles heel can be identified in the process of Arab industrialisation, both civilian and military, then that is the present condition of indigenous scientific research and development. There are four ways of measuring Arab R&D effort. These are based on: published scientific and academic effort: number and scope of R&D centres: number of personnel employed in R&D: and size of funds allocated to R&D. Ironically, the weakness of the Arab R&D effort is exacerbated by the general lack of research on the subject itself. But the few pioneering studies that do exist give an idea of the state of this activity, though still less is known about funding. (These comments apply even more forcefully to the defence sector.)

In one of the few attempts to assess Arab scientific efforts through analysing published output, Antoine Zahlan identified in 1979 33 scientific journals in Arabic and articles by 1,323 Arab scientists appearing worldwide. Of the Arab periodicals, only two related to chemistry and engineering sciences.[20] He also noted that two-thirds of such publications were brought out in Egypt alone – revealing the marked lack in other Arab countries – and that the majority of their contents were linked to empirical findings and experimentation rather than theoretical studies. Furthermore, the actual published output of 1,323 scientists in 1976 represented a mere 10 per cent of the 8–14,000 articles that should, by international standards, have been written by the 7,000 scientists then involved in all types of R&D throughout the Arab world.[21] This compares poorly with the 3,291 Israelis who published scientific articles in the same year, and points to a level of Arab scientific output only 40 per cent of that of Israel.

The last comparison is particularly revealing, as it came despite the enormous disparity in population sizes and the astronomic growth in Arab GNP and in the number of universities and graduates in that period. Indeed, although Arab GNP jumped by a factor of 5.6 to over $140bn between 1967 and 1973, during which time the number of universities rose from 35 to 48 and of their graduates from 300,000 to 850,000, published scientific output stood still at 40 per cent of Israeli levels.[22] Taking the contrasting populations into account, this in effect meant a per capita academic output one-hundredth that of Israel's – better than most LDCs but even further behind than the AICs.[23] Reviewing the Arab situation, Zahlan concluded that, 'the gap between what is actually produced and what could be expected of the manpower engaged in R&D and of the employees and academics in existing institutions, is as wide as can be.' Unfortunately, there is no comparable study using up to date figures, but neither is there any evidence to counter the preceding remarks.[24]

A more recent study provides a broad survey of the number and type of Arab R&D centres operating in the 1980s. Using several criteria, it identifies a basic (though admittedly incomplete) list of

204 centres, departments, and projects, that were running 248 programmes in 1985. There were another 65 centres affiliated to Arab universities, of which only one was dedicated wholly to R&D and 35 partly involved, bringing the total of scientific programmes of all types to 284 in that year.[25]

In assessing these figures, modest as they are, several features stand out that further stress the inadequacy of Arab R&D effort. First, only four out of 144 dedicated centres had a research staff of 200 or over and seven had staffs of 100–200, while 20 employed 30–99 researchers, and a full 113 had less than 30.[26] Secondly, the 284 R&D programmes underway in 1985 included a combined total of 75 for the following fields: nuclear energy, electricity, petroleum and petrochemicals, chemical industries, remote sensors, materials and metallurgy, basic and computer sciences, space and astronomy, information technology, and electronics.[27] Even allowing for the fact that this range includes a number of fields only remotely connected to manufacturing industry (or the defence sector), it represents barely one-quarter of the total number of scientific programmes. A final feature is that, with the understandable exception of agriculture, irrigation, fishing and marine sciences, health/food/nutrition, and applied economics, all the remaining areas of research are undertaken in only between three and ten of the 21 Arab states. This is not entirely surprising, given the distribution of population and resources, but it does reveal that certain countries are particularly behind even by modest Arab standards, and that only a minority of some three to five states may be equipped in R&D terms for industrial development.

Assessing the number of personnel actually involved in Arab R&D is an especially complicated task. A couple of UNESCO studies in the mid-1970s showed the total numbers involved in R&D in 1973 to be a mere 14,265 scientists and engineers, out of a total number estimated at 751,563 in 1973.[28] However, as Zahlan has noted, the former figure includes a large number of part-time researchers, while the latter simply considers all graduates of institutes of higher education in Egypt (593,294) as 'scientists and engineers'. He estimated full-time R&D personnel in Egypt at 3,000 rather than 10,655, and set the Arab total at 7,000 researchers in all fields.[29] Zahlan also observed that some 26.6 per cent of all 347,800 PhD-holders in the USA in 1977 – and 43.7 per cent of the 276,900 PhD-holders in sciences and engineering – were engaged in research: in contrast, figures for Arab higher education graduates at best show a proportion of 5.7 per cent engaged in R&D, or a more likely 2.8 per cent.

The study referred to earlier provides a more up-to-date and more accurate, overview of R&D personnel in the Arab region. It calculates that in 1985 there were 8,014 PhD- or MSc-holders working in scientific institutions other than universities.[30] This works out at roughly 1.7 R&D personnel per 10,000 members of the general workforce (though the actual ratio varies from below and up to six in different Arab states), compared to ratios in the AICs of 66 (USA), 99 (USSR), 48 (West Germany), 58 (Japan), 36 (United Kingdom), and 39 (France).[31] If 10 per cent of university academics are assumed to be engaged in research too, raising the total to 12,712, then the Arab ratio reaches 2.7.[32] These recent statistics are evidently significantly below those mentioned for the 1970s, but, besides the fact that they are probably far more discriminate and accurate, even the higher estimates would still point to a seriously lacking Arab R&D effort. In the meantime, the expansion of education and certain economic sectors (such as services) continues to draw large numbers of professionals away from R&D.

The picture becomes even worse for Arab industrial development when one takes into account the breakdown by category of R&D. According to the same study, of the 8,014 researchers employed in non-university settings in 1985, only 31.5 per cent were engaged in any of the fields mentioned above that could conceivably be connected to industrial activities.[33] Of this proportion, barely half (16.1 per cent) had any potentially direct industrial connection, while the share of advanced technologies (electronics, information, computer, and biotechnology) was a mere 0.9 per cent. Expressed the other way around: although agricultural, medical, marine and animal, and other non-industrial sciences have a major significance for Arab economies and societies, industry simply cannot become 'the major potential leading sector of growth and structural change' without a far greater R&D allocation.[34]

The final indicator of the status of R&D in the Arab countries is the amount of funding it receives. In 1984, the latest year for which comprehensive statistics are available, the dedicated research centres across the Arab world (but excluding universities) received a total of $369.5m in their budgets. This represented an investment equal to under one-tenth of one per cent of aggregated Arab GDP in that year.[35] Even when 5–10 per cent of all funds allocated to universities are also added, on the assumption that they devote part of their effort to R&D, its share reaches only 0.13–0.17 per cent of GDP.[36] (Universities, incidentally, receive 0.71 per cent of GDP, indicating a combined expenditure on higher education and R&D of 0.8 per cent of GDP.) These levels of funding are roughly half the Third World average of 0.3 per cent of GDP spent on R&D, and less than one-sixteenth of the average 2.5 per cent in the AICs.[37] That said, the study supervised by Qasim notes that the allocation of R&D funds within the recipient institutes is inadequate: little money is actually spent on R&D at universities, while at the dedicated centres more outlay goes to pay salaries than to cover operating costs.[38]

Scientific and Technical Industrial Manpower

The final prerequisite for successful industrial development (both military and civilian) is the availability of trained manpower, in adequate numbers and with the appropriate qualifications. Here the Arab situation is mixed, with clear advantages and disadvantages. There can be little doubt that the strongest point of all is the existence of a large and growing manpower pool. The projected Arab industrial workforce for the year 2000 includes: 3.3–4.4 million engineers, technicians, and administrators; 14.9–19.9 million skilled labourers; and 4.5–6.1 million unskilled workers.[39]

Central to achieving this projection is the role of a large number of institutes of higher education – universities, polytechnics, and technical colleges – and vocational and technical secondary schools. This sector has produced hundreds of thousands of graduates with scientific or technical training in the past two decades, and will produce many more by the end of the twentieth century. At least on paper, Arab industries should have no difficulty in finding their manpower needs of all types and levels.

In reality, however, the Arab technical manpower situation suffers from several inherent flaws. Among the most important of these is the major lack of technicians. In effect the latter form the backbone of any industrial (indeed of virtually any economic) endeavour, and so they play a key role in development. This is particularly applicable in the case of arms production, which 'requires machine-shop skills such as are found in tool-makers, machinists, fitters, drillers, borers, lathe operators, forgers, reamers, press workers, heat treatment specialists and so on'.[40] Furthermore, not only does this shortage adversely affect production, it also seriously undermines critical support services such as maintenance and repairs both in industry and in user sectors.

Given their importance, some experts consider that there should be a ratio of up to 100 technicians to every engineer in industry, in order to achieve full results in terms of volume and quality (as well as overall efficiency) of production.[41] Yet the Arab countries do not approach such levels even remotely, neither in relative nor in absolute terms. The problem is equally evident in the systems used to provide technical manpower. In most industrialised nations, the bulk of technicians are produced not by formal educational establishments, such as technical and vocational schools, but through schemes designed to provide on-the-job training or apprenticeship. The latter may be conducted within the industry, by correspondence, or in collaboration between two or more of the main sectors concerned: industry, educational system, and government.

Such schemes are virtually non-existent in the Arab countries, however. Moreover, although a preferred technicians-to-engineers ratio of at least 3:1 should be attained even within the formal education system (vocational/technical schools versus university and polytechnic graduates), the reverse applies in the Arab case. Even a country geared toward exporting skilled manpower, such as Jordan, had only some 7,000 students enrolled at technical training schools in the mid-1980s, in comparison to an estimated 89,000 Jordanians at local and foreign universities.[42] The lack of technical training is also reflected in the disproportion between registration at vocational/technical schools and general secondary school enrollment. Thus while the ratio in the AICs in the early 1980s stood at 21.3 per cent, it was only 11.4 per cent in the Arab countries. More to the point, the Arab level was 3.4 percentage points below the world level, which includes all other developing countries.[43]

A second manpower problem is the imbalance between different types of engineer. Not only does the Arab educational system concentrate on producing high degree cadres at the expense of technicians and other skilled workers, but it also tends to produce more design engineers than production or maintenance engineers.[44] Thus production is again undermined by the lack of dedicated professionals, a disproportionate number of whom are instead directed to white-collar jobs or supervisory functions.

Ironically, despite the imbalance among engineers in favour of non-productive functions, there is a critical shortage of industrial managers of all types. This is a problem of major magnitude, almost on a par with that of the lack of technicians, as it is a principal contributor to low productivity and under-utilisation of capacity and resources – in other words, to a general and substantial loss of efficiency. The detrimental effect of the shortage of managerial skills is also reflected in poor marketing, among other areas, further reducing the commercial feasibility of local defence production. Although the situation is somewhat better in the arms industry owing to the high levels of capital and human investment and of opportunity/substitution cost, Arab average productivity remains very low, at only $2,803 in 1986.[45] The general problem is further demonstrated by major Arab reliance on foreign managerial skills, which account for one-third of total Japanese and American exports of consultancy services.[46] This is underscored by the fact that the Arab region has only some 3,000 experts dedicated to managerial development and research, although based on population figures there should be an estimated two million managerial experts, according to one leading Arab analyst.[47]

A final comment on the Arab technical manpower situation is that, parallel to the functional imbalance, there is a problem of geographical maldistribution. This is a result of local historical circumstances, and may be both natural and beneficial in the case of industrialising countries such as Egypt, but has led to a mismatch of resources and capabilities in other cases. Thus a country like Jordan is manpower-rich but lacks the ability to establish a major local industry, whereas most Gulf states are severely short of skilled human resources though well-endowed with the wealth to industrialise. The central implication here is that there is significant potential for the exchange of personnel and services between the Arab states, both in the military and the civilian sector, though this will not resolve the broader problems of lack of technicians, production engineers, and managers.

Conclusion: Arab Military Industrial Potential

Although vital, the indicators discussed above – civilian industrial base, R&D base (including scientific manpower), and technical manpower – remain general and are insufficient to determine the actual or potential ability of a given country to establish and sustain an effective local defence industry. A more precise set of criteria need to be defined.

In developing a more refined approach, Herbert Wulf starts with the same general requirements posed above, and then leads into other issues. Essentially, he suggests that questions must be posed relating to the following three areas: the degree of diversification of the country's industry and its level of competence in terms of skilled manpower and R&D; ability to keep production costs low or achieve exports, and the availability of local financing; and control over production technology and access to foreign sources, with associated implications for political relations with suppliers, costs of acquisition, and indigenous capabilities.[48]

Once again, these remain relatively general, though critical, determinants of military industrialisation capability. Wulf therefore goes on to define what he calls the *potential arms production base*. This is a measure of the actual or potential ability of a given country to produce military hardware locally or to raise the level of indigenous content. In fact, the potential arms production base divides into two categories. First is the industrial base, which consists of six key industries: iron and steel, non-ferrous metals, metal products, electrical machinery, non-electrical machinery, and automotive or transport equipment.[49] The second category – the manpower base – consists of the employees and scientists engaged in those six industries or in related R&D.

Viewed in these terms, on the basis of the data presented in this chapter, a tentative conclusion can be drawn that the Arab countries are not particularly well-equipped for large-scale military industrialisation. From one Arab state to another, the level of capability probably varies from low to medium, although specific countries (such as Egypt) enjoy some advantages in certain sectors. There is evidently massive need for development in all areas – industrial base, R&D, and manpower – before defence production can be ensured that is both commercially and economically viable and reasonably self-sustaining technically and technologically. Yet again, progress in this regard is very much part and parcel of the broader problem of formulating improved industrial, technological, and scientific policies and of implementing structural and organisational reforms.

PART FOUR

Determinants and Strategy of Arab Military Industrialisation

11

Issues and Determinants of Future Military Industrialisation

Aims of Defence Production

The aims of Arab military industrialisation are, and in the foreseeable future will continue to be, set by the small circle of central, top decision-makers in each country. There has been little public input into the process at any level – especially that of policy – such as the recommendations of experts and academics. Taking this as a starting condition, the question then devolves on to the decision-makers to define the main aims of local defence production. This is an absolutely necessary precondition for the successful and efficient functioning of any enterprise, and particularly so in the case of the defence industry, which is characterised by security sensitivity and high costs. Definition of aims also involves defining assumptions and setting criteria, in order subsequently to evaluate performance.

There are a variety of possible aims for defence production, which need not be mutually exclusive. One is to secure military needs and ensure the flow of combat supplies at all times. A second is to secure those needs at the lowest possible cost – in other words, to reduce defence spending. A third is to assist development through stimulating the industrial and scientific base, within an inward-looking strategy. A fourth is to transform the sector into an export-oriented one geared to making profits as its principal function.

Choice of aim affects industrial strategy. In the first case, for example, Amin Huweidi argues that the technological dependence caused by local arms production is actually more harmful than conventional dependence for arms supplies. He therefore suggests that if the purpose is to ensure basic military needs, then diversification of external sources is more effective and far less costly and debilitating than local production.[1] This simply goes to show that defining a specific objective may lead to the conclusion that a local arms industry is not the answer. It also shows the need for a more sophisticated and varied approach to military industrialisation that accords differing levels of priority and that establishes a number of horizontal and vertical linkages in order to achieve better results in more areas for lower costs.

At present, Arab officials still offer relatively obvious targets at the overall, or strategic, level. As late as 1988, Abu Ghazalah for one was asserting that 'one day all the Arabs should depend entirely on Arab-built military equipment'.[2] A Saudi official meanwhile emphasised the importance of achieving the complementarity of industrialisation in order to preserve national security. What is not clear is whether such statements reflect a specific intention, and indicate a particular strategy. Nor is it clear whether there is a process of internal debate at top levels, that leads to decisions being taken that relate to the general strategy and aims of local defence production. There is, of course, clarity about the initial reasons for setting up a defence industry, as the case of Iraq during the Iran–Iraq war shows. But it is not at all evident that the further implications and broader possibilities of each option are fully discussed. This therefore remains a primary task of Arab decision-makers and commentators.

Needs

General aims are, of course, a response to broad needs, and, once defined they impose their own practical needs and requirements in turn. Thus there are two main levels of needs: overall national needs, such as preserving national security and independence, ensuring combat supplies for physical defence, and improving the economics of defence spending; and the technical, organisational, and financial means for achieving set objectives. The former are an essential part of stating national aims and formulating strategies. The latter deal with implementation, and effectively divide into two sub-levels: systemic/operational requirements, and technical/industrial requirements.

Systemic/operational needs refer to the types of hardware and service needed locally, and to the general civilian industrial and scientific base and R&D network that successful defence production requires. In the first place, the national leadership identifies its defence needs according to its perceptions of the external and the internal environment. An unofficial study defines three types of war that Arab states must prepare for: counter-insurgency, regional inter-state conflict, and 'popular' or regular war against advanced, high-tech powers.[3] The first of these would require low-to-medium technology levels and smaller expenditure. So-called 'popular' war would require broadly similar technology but greater amount of hardware, while all other forms of conflict would demand increasingly high-tech weapons. Each choice would, moreover, involve varying types and degrees of repair, maintenance, and overhaul services, as well as training, construction, and other follow-on support. It would be up to each leadership to determine how to obtain each need – through import or local effort. This choice will tend to coincide with the country's industrial capabilities and economic strategy: an inward-looking import-substituting industrialisation (ISI) strategy will seek to produce locally a wider variety of items of higher technological level; whereas an export-oriented approach may focus on specific areas of production and lower technology levels. Thus broad operational needs will help to define the industrial system required.

As to technical/industrial needs, these relate to items required for actual production, such as process and design technology, manpower skills, and metallurgical or chemical sciences. The study referred to above cites two types of technology required, which it defines as: central and supporting. The first category includes main systems (design) plus supporting technologies, like electronics and computers or metallurgy and chemicals; while the second refers to engineering sciences, applied physics and the like. Both, however, are essential to the process of military industrialization. Other technical/industrial needs are raw materials and primaries, finished or semi-finished components, and production machinery, jigs, and tools. Developing manpower skills is an important requirement, whether for workers, technicians, engineers, managers, or marketing agents. Exactly which items are needed most, and how to acquire them, will vary from one country to another.

Determinants

In seeking to formulate a strategy for military industrialisation into the 1990s, the Arab arms producers must secure a number of determinants, without which any such ventures will fail. The main areas in need of development are: technology transfer, R&D, information base, funding, and civilian industry.

Technology Transfer

At the end of the day, technology remains the single most important driving element behind any industrial activity, virtually by definition. From an aircraft engine or tank gun down to the smallest screw used in riveting an aluminium panel, the producer must possess knowledge both of overall design and of the methods and materials required to turn a blueprint into physical reality. Thus, however technology is obtained, it is crucial for the development of all system designs (hardware) and manufacturing processes. A primary task for Arab defence industries is therefore to ensure the acquisition of such know-how, of which transfer from abroad is a foremost method.

The last point cannot be stressed enough. It may just have been possible, by a wide stretch of the imagination, for LDCs to embark in the 1960s on the independent development of military technology, as indicated by India's *Marut* HF-24 fighter aircraft or Egypt's HA-300 fighter and *Zafir/Qahir* ballistic missile programmes (though in fact all such efforts relied in varying degrees on imported designs, parts, and/or experts). But, by the 1990s, the massive strides in both systems and process technology have undeniably left the LDCs far behind the AICs. With an evolving world economic system marked by fierce commercial competition, a number of specialised technologies have become ever more concentrated in the hands of a very few countries or even companies. Examples are engines and composite materials, which are jealously guarded secrets. For various reasons a variety of minor items are also produced in only a small number of places – aircraft ejection seats are an example. Furthermore, most modern technologies undergo a constant cycle of obsolescence, due both to inherent traits and to the continuing revolution within them, which means that soon after an LDC has acquired a certain level of know-how it becomes outdated and less competitive in the market and on the battlefield alike.

Thus attempts at complete self-sufficiency in weapons production are self-deluding and extremely costly. Measured by the investment of time and money in it, moreover, local development may even backfire, because military and economic strategies have been based on particular technologies that have since been superseded. And, as authors such as Wulf and Huweidi have noted, the need for technology only binds LDCs to a dependence on the AICs that is even deeper and more systemic than it was for arms imports. Furthermore, even attempts to develop specific vital technologies, like engine design, and production indigenously are doomed to be protracted and almost inevitably futile. This has been the experience of India's *Arjun* tank and LCA fighter aircraft, both of which have been repeatedly delayed by difficulties in producing competent power plants locally. Even an industrial power like Japan imports certain types of technology: after years of development, its locally-designed engine can only produce one-seventh the power of the P&W 1100 that drives the American F-16.[4]

The purpose of these comments is not, however, to inspire defeatism or to suggest that all efforts at local, independent technological development are futile and that continued dependence on foreign technology transfer is inevitable. Rather, the aim is to generate awareness of the issues and difficulties, and so to encourage a varied approach that is most appropriate to both needs and realities. Thus, accepting that Arab arms producers must look to the outside for at least part of their technological requirements, the question is how best to ensure transfer in defined areas and to specified degrees in each case.

Again, it is beyond the competence of this book to provide a detailed list of specific requirements in answer. Instead, the aim here is to stress a number of guidelines to be followed in any strategy for achieving technology transfer and benefiting the most from it. The first task is to pose a question: is technology transfer in a given field being reflected in increased independence in manufacturing or design ability, or in new patents or markedly improved products? The real purpose of transfer is ultimately the acquisition of skills as well as knowledge, and so the answer will reveal how much use is being made of opportunities and where the obstacles are. Moreover, the emphasis on skills acquisition encourages an outlook that seeks an opportunity to achieve technology transfer in every type of economic activity or exchange; even if there is heavy reliance on foreign workers and experts, the relevant Arab authorities can adopt a deliberate policy of training their nationals and encouraging them to learn skills on the job. This is especially important in countries like Iraq and Saudi Arabia which have resorted overwhelmingly to expatriates even for basic tasks such as civilian construction.

A second guideline is to stress the development of process, rather than system technology as Mary Kaldor suggests.[5] She notes that there is too much emphasis on product innovation – using technology to improve the hardware itself – rather than on refining production processes in order to reduce costs. 'Combat production' costs could also be lowered by relying more on equipment such as PGMs, and less on major weapon systems, such as armoured vehicles and aircraft, that may soon become obsolete and that requires design and production capabilities in a large number of subsidiary and associated areas. An important added advantage of this approach would be to develop an Arab ability to approach any manufacturing task in general.

By mastering process technology, Arab producers can expand their industrial and scientific base and so benefit both the military and the civilian sector, and they will be better placed to apply and adapt specific design blueprints (whatever their source). It is this ability that enabled Israel and South Africa, for example, to survive arms embargoes and develop dynamic local defence industries. Put differently, if transfer from abroad is unavoidable, and if military industry will always be costly, then it is best to shift the allocation of resources towards acquiring those capabilities that are most enduring and self-sustaining.

Thirdly, technological requirements should be distinguished according to type and area, in order to assign different priorities and develop varying means of acquisition. For example, Arab defence industries might continue to rely predominantly on foreign licences and transfers in the production of major weapon systems such as aircraft, but seek to develop an indigenous capability in other branches, such as electronics, computers, optronics, and robotics. True, these are capital- and skill-intensive areas of technology, that, moreover, necessitate recourse to foreign sources, but the costs of acquiring them on the national level are substantially less than those incurred in setting up an aerospace industry.

Furthermore, high-tech products are the cutting edge of the technological revolution, and therefore of the modern battlefield and the world economy. They are, moreover, the most expensive components by far in the production of any major weapon system and a basic element in any product modification or improvement, and are particularly amenable to utilisation in non-military applications, thus opening up greater prospects for civilian industry spin-offs. This also means that the cost of acquiring them might be spread more broadly.

In all cases, foreign countries and companies remain the main source of technology (both systems and processes). Direct purchase of technology packages or licensed production are still the most obvious, and probably the most cost-effective means of achieving technology transfer. However, the fact that the AICs are the principal, almost the sole source of advanced technology does not mean that LDCs cannot loosen their grasp. What is required is the technological breakthrough strategy implied in the preceding paragraph. This could mean focusing effort on particular areas in order to master their technology and so create a starting base for further, independent development.

In order to do so, any number of means may be used to circumvent AIC controls and restrictions. The most straightforward is to use any existing commercial or political advantages in order to purchase needed technology or to persuade foreign states and companies to allow transfer in specific areas. An example is the success of India in obtaining American permission for the transfer of primary composite and flight-control technologies for its light combat aircraft (LCA) programme.[6] Naturally this is not possible unless the recipient has clear industrial capabilities, commercial or strategic importance for the donor, and the economic sense and political will to insist on such transfers.

Since technological breakthrough is based on acquiring a 'critical mass' which permits independent development from thereon, another method to attain it is by circumventing government controls. Unofficial or plainly illegal transactions have been the basis for local production of several major weapon systems in Israel and South Africa, not least of which was the famed *Mirage* III incident in which a Swiss engineer was arrested for selling blueprints to Israel. More recently, an Egyptian effort was uncovered in the USA in 1988, aimed at smuggling prohibited technology relating to guidance systems design and carbon-carbon production, probably for the Egyptian ballistic missile programme. And in September 1989, Western media publicised details of an extensive Iraqi commercial and financial network set-up, among other purposes to acquire military technology. This approach coincides with an inward-looking industrial strategy, which stresses the need to learn foreign technologies and adapt them.

The essential element in achieving technological breakthrough is not, however, finding ways and sources for acquiring blueprints or basic knowledge. The existence of a large number of private companies in the AICs engaged in high-tech and defence activities, and the growing world-wide competition between arms producers, suggests that LDCs with the financial means are able to obtain many of their requirements on a private basis. What is of paramount importance, then, is the creation of a local ability to absorb and utilise acquired technology, eventually to expand and develop it. This is dependent on establishing an appropriate scientific base in the recipient country, which in turn feeds a dynamic industrial base.

R&D

Central to achieving effective technology transfer, then, is the R&D effort. This is one of the weakest elements in the Arab military industrial effort, as argued earlier. Lack of proper R&D slows down the transition from one level of production to another (such as from assembly to parts production), affects quality of products (and therefore reduces competitiveness and combat effectiveness), and inhibits the emergence of a local ability to modify or design equipment. Without an adequate R&D base new technology cannot be absorbed, nor will the accumulation of skills lead to improved production or independent capabilities. A protracted R&D effort has been at the heart of India's ability to develop a wide-ranging defence industry that is scheduled to produce indigenously-designed tanks, combat aircraft, and major naval vessels (including an aircraft carrier) in the 1990s. Along with the advantages of an energetic civilian industry, it was also the basis for the success of Brazil's low-to-medium technology military hardware exports in the 1980s – in contrast to the difficulties Egypt faces in marketing its own, comparable goods.

At present, Arab R&D levels are seriously below world averages in terms both of funding and manpower. According to the study supervised by Qasim, Arab expenditure on dedicated R&D centres ($369.5m) stood at 0.095 per cent of aggregate GDP in 1984. Adding a generous 5–10 per cent for expenditure on universities for R&D work done there, the final share rises to 0.13–0.17 per cent at its highest ($507.2–644.9m).[7] (For an idea of scale, this meant a per capita R&D share of $2.85–3.62 annually, for a total Arab population of 178 million.) This is not only a fraction of the amount spent on R&D in the industrial North (2.5 per cent of GDP), but at best roughly half the average in the developing South (0.3 per cent).[8] For further contrast, Iran currently spends 0.5 per cent of its GDP on R&D, and plans to raise that to 2 per cent, while South Korea already devotes 2 per cent (to rise to 3 per cent by 1992). In India, an over-populated and poor country with a per capita GDP less than one-quarter that of Egypt, national R&D receives 0.5 per cent of GDP.[9] In manpower terms, the picture is equally bleak. Compared with a normal ratio required for proper R&D in all fields of 2,000 researchers per 1 million inhabitants, it is estimated that the Arab states will have an aggregate average of only 250–450 by the year 2000.[10] Western Europe is expected to have 2,200–2,500 by that time.

The R&D problem is not simply one of budgets and manpower, though these are essential. It is also one of organisation and structure. Little precise information is available on Arab military R&D, but it is generally limited to relatively small bureaux affiliated to specific factories and companies. Only in the case of the Egyptian ballistic missile and fighter programmes of the 1960s was there an evident attempt to establish a large and diversified R&D infrastructure dedicated to the task. Iraq may have been developing its R&D effort before 1990, as indicated both by its employment of foreign experts and acquisition of technology, and by the obvious modifications that were introduced to a wide range of weapons and equipment. However, it remains unclear whether there is a co-ordinated, industry-wide scientific programme in any Arab arms-producing country (besides the nuclear field in pre-1991 Iraq). There is almost definitely little systematic co-operation with academic centres outside the defence industry itself, whether at the universities or in the civilian public and private sectors.

This contrasts sharply, on every count, with the situation in other Third World countries. A major element in the success of the Taiwanese defence industry (and in the South Korean), for example, has been the formation of a central, government-run institute for science and technology (CIST), which already employed 2,000 professionals by the mid-1970s.[11] Of that number, 50 per cent came from universities and private industry. In India, R&D centres employ 25,000 people in the defence sector alone, while in Iran the involvement of a large number of educational establishments (including vocational and technical schools) and private workshops played a considerable role in expanding local defence production during the Iran-Iraq war (although the long-term viability and effectiveness of the Iranian sector may be questioned).

Thus the establishing of a local R&D capability is a foremost task for the Arab arms producers (as it is more generally of all the Arab countries). This can probably be best achieved by forming a central co-odinating body in each case, with branch-specific complexes that both undertake research and direct subsidiary programmes within separate factories and companies involved in a particular field, such as aerospace or electronics. In all cases interaction should be as direct as possible with the users – the armed forces or foreign clients. These bodies might be primarily set up on a country-basis, with higher co-ordinating committees at the pan-Arab scale, though specific research areas and programme might be managed more effectively through inter-Arab co-operation and division of functions from the outset.

It is also essential to provide adequate funding, not only to state-owned enterprises, but to academic institutions, dedicated non-government centres, and private companies engaged in defence-related work. This could be channelled through the central co-ordinating body, and solicit private as well public funds. Funding in particular could also be an area for inter-Arab co-operation.

Information Base

As important a task is to establish an extensive Arab information base. This is an essential first step in defining needs and capabilities – scientific, technical, industrial, financial, and military. R&D activity can then be directed more economically and effectively towards the areas of greatest need, utilising existing resources more fully. Information is also essential for rationalising the whole gamut of industrial activities: from purchasing raw materials or shipping them (using Arab transport), through supplying components and semi-finished products, to assembling and marketing the end-products.

In practice this means setting up dedicated centres at both the national (country) and inter-Arab levels, that compile all relevant materials in various languages, link up with relevant academic and scientific bodies both within and outside the Arab World, host seminars and study projects, and ensure full dissemination of data and other materials. For military R&D specifically, this is more likely to occur on a country basis, or even within the main branch of the defence industry itself, though that by no means denies the urgent need for a generalised, inter-Arab, scientific information base. In general, the importance of information (and its associated technologies and hardware) can hardly be overstated. It is the basis of most economic and military activity anywhere and is fast becoming a major component of world trade. Increasingly, it is those LDCs that are unable to keep pace in the information revolution that are falling further behind in every respect, and this is a threat that confronts the Arabs all too seriously.

Implicit here is the need for the formation of a multi-disciplinary scientific and technological network spanning the Arab countries. This is partly a formal organisation – that is, the official channels of contact established between the largest number possible of R&D-engaged centres. Furthermore, the concept of networking coincides with the need to rationalise the Arab computer sector: both to develop an indigenous software and hardware design and production capability, and to standardise criteria and co-ordinate imports and development.

The informal network is as important – it consists of the host of direct and personal contacts between researchers and centres, backed by a vitalisation of Arab scientific publishing. The last is an essential means of disseminating the results of R&D, and more generally of encouraging co-operation and cross-fertilization between Arab countries and of generating a 'scientific/industrial culture'. And again, although the results of military R&D are unlikely to be published, it will benefit from the provision of a public base of reference. None of this can be achieved, however, unless adequate funding is provided (especially for modern, computerised information bases) and the freedom of access to materials and information from various sources is assured.

Funding

The importance of funding for military industrialisation, as for any other economic activity, hardly needs confirming. It has already been mentioned in this section in connection with the transfer of technology and R&D. As a determinant of defence production in general, however, funding gains even greater importance. Policies for financing local arms industries will vary according to the overall economic and industrial policy in each country, but some basic observations can be made.

First, although the issue of securing foreign investment may be regarded more or less favourably in some countries, the experience of most Third World arms producers is that defence production is far less likely to receive external funding than any other activity. This even applies to countries like South Korea and Singapore, which have adopted an industrial strategy that effectively turns parts of their defence industries into subsidiaries of Western 'mother' companies. Such foreign investment as does occur most often takes place in enterprises geared mainly to civilian production, but which also produce some items (such as engines and tyres) for the defence sector. In general, foreign funding is least attractive (or desirable, in any case), but if it is seen to be feasible then the ability of each country to obtain it depends largely on the dynamism of its industry and its ability to generate offsets or other advantages for the rest of the economy. The benefits will be lowest for a country that has to resort to borrowing – or, arguably, to increased sales or barters of ready surpluses such as oil – in order to secure loans or invite partnerships.

Secondly, although it is not being argued here that there is an intrinsic advantage to raising the share of private industry, this can nonetheless be a valuable source of additional funding. Private participation is less likely in major weapon programmes, owing to the massive investment required, but its cumulative role may be substantial in associated ventures such as high-tech areas (computers, electronics, or optronics). There the starting capital is lower while opportunities for personnel recruitment and advancement are higher, and diversification into civilian production is both possible and feasible.

The defence sector is, moreover, in constant need of a large variety of primary materials and components from the civilian feeder industry, which offers another area for private participation. Indeed, since private investment is most likely in the civilian sector, which can generate large cumulative values, its benefits may well be reflected in higher quality goods for the defence industry and greater productivity and contribution to the national economy. However, the private industry is unlikely to expand in the context of restrictive commercial and financial regulations, not only because of limitations on profits or capital transfers, but also through difficulties in arranging imports of needed materials and exports of finished goods, problems with transport and marketing, and the inadequate role of the banking sector.

With these considerations in mind, the issue is how to increase funding for Arab defence production in a manner that ensures utilising it most effectively and that spurs broader participation by all sectors. The question is evidently only relevant if private participation is assumed. Otherwise defence production would remain a wholly state-owned affair; in that case the capital is either provided, or underwritten, by the government, which, moreover, effects disbursement through its conventional channels or an industrial authority. The main exception to this statement would be a joint Arab effort, such as the AOI of 1975, which would require not only a senior financial body, but also a structure that sets principles for disbursement, repayment, and settlement in case of dissolution.

Both the Saudi SIDF and the Turkish DIPF offer examples of the sort of specialised fund that need to be established in order to stimulate private participation in the arms industry. However, the first has a broader mission that goes beyond the defence sector. More importantly, the Saudi MIC – the body entrusted with the task of co-ordinating state funding with private industrial companies – does not offer a guiding strategy, nor does it define projects and needs or award contracts. Thus there is little co-ordination between the public and the private side of the defence sector, and next to no overall strategy for either to implement. Conversely, the DIPF in Turkey is closely integrated with Defence Industry Support and Development Agency (DIDA) – the body that both sets industrial strategy in the defence sector and oversees implementation – and so makes use of its funds more effectively. The Turkish government is moreover committed by law to set aside $500m annually for investment in the defence industry, which assures the private sector of continuity and so makes its own commitment feasible.

The conclusion is that Arab defence industries would benefit from state-directed funds working in close co-ordination with higher policy-making bodies. These funds could be either wholly provided by the government or partly financed by the private sector (industrialists and financiers). Saudi Arabia is obviously best placed to implement this approach, given its wealth and encouragement of the private sector, though it is otherwise the least motivated or equipped generally. Egypt lacks the ready funds; it might generate greater private industrial investment by removing certain restrictions, but would still need substantial 'seed' capital from the state before a mixed private–public fund could be set up. And although the situation has changed radically since 1990, Iraq may have felt previously that it had enough state revenue to preclude the need to involve the private sector, which was in any case severely restricted. If it is to achieve industrial revitalisation and cope with post-war reconstruction and development then it will need to generate massive resources, part of which might come from private participation.

The example of the AOI, moreover, shows that it is possible, given adequate controls and criteria, to establish funds on an inter-Arab basis that have the partial task of assisting private participation in the defence sector. This was directed to foreign investment in the case of the AOI, as it turned out, with four joint ventures being formed with American and British companies, but the same structure was theoretically open to local private capital. A similar arrangement could apply to Arab support for a country-based fund in which, say, Saudi money could be channelled through an Egyptian equivalent of Turkey's DIDA in order to finance specific private companies engaged in developing equipment of use to the Saudi defence industry or armed forces.

This is a function that could alternatively be managed through multilateral, non-profit, Arab development institutions. Such specialised collective agencies could also fund particular branches of local defence industries – such as shipyards or computers – that fall within their area of specialisation. Adequate financial and technical guarantees would, of course, be necessary in all cases, just as they would for any inter-Arab fund based solely on state participation (which points once again to the importance of an information network).

Civilian Industry

As discussed in previous chapters, the level of development of civilian industry is an important determinant of defence production. It acts as a 'feeder' for primary materials and components used in producing military hardware, and may develop specialised areas, such as electronics, chemicals, or metallurgy, that are needed in both the military and the civilian sector. It also trains industrial manpower and contributes to the prevalence of a general scientific culture, and provides experience in fields such as management and marketing. The presence of a dynamic and capable civilian industry is, moreover, an essential condition if the national economy is to achieve maximal side-benefits and spin-offs from the defence sector.

Yet civilian industry is precisely an area of marked weakness or shortcomings in the Arab countries. This is evident in the modest share of manufacturing activities in Arab GDP – particularly when these are reduced to the production of metals, machinery, and the like – as detailed previously. It is also evident in inadequate manpower policies and the imbalance in technical training. Specifically, Arab educational institutions are preparing a large number of cadres with high degrees, but relatively few technicians and skilled workers.[12] With a total university population of four million (including some 320,000 abroad), the actual capacity of Arab technical training centres represented the merest fraction.[13] And even among university scientific graduates, far too many are design engineers, instead of being production or maintenance engineers.[14] Yet, paradoxically, there is a surplus of unemployed technicians, or of engineers filling the positions of technicians, because of the structural and 'contextual' flaws in Arab industrial and educational policies: demand for technicians is low because of high reliance on imported goods and services, and there is virtually no co-ordination between industry, educational system and relevant government agencies in the training and distribution of personnel.

Nor is Arab industry helped by the battery of regulations affecting trade, especially inter-Arab commerce. In addition to import and export controls and monetary restrictions, there are obstacles such as the lack of adequate transport and marketing networks.[15] The absence or weakness of collective, inter-Arab co-ordinating bodies further undermines industrialisation and trade, which in turn prevents Arab industries from achieving economies of scale, among other negative consequences. These problems are, of course, over and beyond the structural flows internal to each Arab country, such as low private participation, inefficient state administration and planning, and distorted approaches to development. This is often reflected in the inadequacy of civilian R&D (though it receives more of national R&D budgets, which are low enough, than the defence sector) and low technology transfer – evident for example in reliance on turnkey projects in all fields including construction.[16] These are, moreover, symptomatic of the absence of guiding policy in key branches (such as electronics or computers).[17]

The implication of this argument is that development of civilian industry should be made part of any general plan for Arab military industrialisation. Otherwise local defence industries will rapidly reach a limiting ceiling in their development potential and suffer under-utilisation of actual or potential capacity. The cost of maintaining them will rise sharply, especially in the light of the spiralling costs of military R&D and the production of modern weapon systems globally.

Undertaking development of the entire civilian industrial sector is beyond the means of any arms industry, of course, but the latter's own development can be structured in a manner designed to achieve maximal benefits for the civilian sector and the greatest economies nationally. (Incidentally, the preceding also reaffirms the necessity of generating bigger financial resources for civilian development by encouraging private involvement, if the task is to be achieved on a national scale.) Conversely, investment in civilian industry can be directed to those areas that are most likely to benefit the defence sector while enjoying high commercial feasibility in civilian and export markets, such as electronics or special metals. This approach would prove more attractive to private and foreign investors (if that were an aim of the government concerned), who are normally reluctant to risk low profitability. More importantly, it would also benefit from the fact that, despite high capital investment, the cycle of obsolescence in civilian manufacturing is far slower, thus making for better cost-effectiveness.[18]

Utilising military and civilian industrial development interdependently can be achieved in two main ways. The first is to use any excess of capacity within the defence industry to produce civilian goods, as has taken place in Egypt. There the NODP and the AOI feed raw materials and semi-finished parts and components to civilian industry, housing, land reclamation, transport and communications, electricity and energy, petroleum and mining, supplies, education and public health.[19] The same applies to the Iraqi defence industry, though details of its civilian activities are not available. Furthermore, rather than use incidental excess capacity, planning within the defence industry could in fact design built-in excess with the express purpose of undertaking civilian production – an approach that would be feasible in Iraq as well as Egypt.

The second method of creating a constructive linkage between the military and the civilian industrial sector is to develop particular areas of activity that are both easier to achieve (in human and financial terms) and most beneficial economically. An example is technical services such as maintenance, repair, and overhaul of equipment. These are skills that the civilian sector can provide at reasonable cost, and that are beneficial to all parts of the economy. A similar logic applies to construction and engineering, which are an additional major component of defence expenditure. Indeed, these areas of support services can make a contribution to local defence activity that is potentially equal to, or even greater than, that of actual arms production. And, although the level of reliance on foreign services varies from one country to another, providing these services locally can offer massive savings on imports and hard currency and expand GDP significantly.

In addition to services, civilian industry could feed the defence sector more effectively by deliberately reorienting or adapting some of its substantial excess of capacity towards specific types of production. This is not to ignore the fact that under-utilisation of capacity is partly due to poor management, structure, and organisation, and that the civilian market would be quite capable of absorbing any increase in production resulting from reform. Nonetheless, the remaining excess of capacity could still be diverted to producing special primary materials or components for the defence industry. Indeed, an excess of capacity could be built into the civilian sector with the express purpose of supplying that industry – the advantage being that larger civilian industrial units can be planned with longer production runs and lower final costs.

A particular area in which a major contribution could be made by the civilian industry, for which it would be profitably expanded, is the provision of the wide range of non-combat military equipment required by any army: tractors, bulldozers, trucks, jeeps, tankers, hoists, workshop machinery and lathes, electrical machinery and generators, bridging equipment, piping, plastic products, and a host of other items. (This is not to mention the obvious, often-referred to electronics sector.)

Industrial and Import Policy

A final determinant of military industrialisation strategy is the policy adopted by arms producers towards industrial development and towards imports or import-substitution. More specifically, what is important is to determine how industrial activity and imports are related to, and affected by, each other. This will ultimately determine the nature of local defence production, its technological level, policy and means of technology transfer (and R&D), overall aims, and relationship with the civilian sector. It will also provide a powerful argument in favour of maximising inter-Arab industrial and commercial co-operation in all its forms.

Several elements form the background to this discussion. First is the fact that the Middle East has the highest ratio of arms imports to total military expenditure among all Third World regions.[20] In itself, this indicates the wide scope and marked desirability of developing local defence production in order, at best, to reduce imports and save hard currency, or at least to alter the composition and direction of imports and utilise spending with better results for GDP. However, the massive scale of Arab arms imports, and the fact that much of them consist of high-tech major weapon systems, equally casts doubt on the practical feasibility of seeking wide-ranging import-substitution. A further point to note is that resort to diversification of foreign arms sources, as an alternative to domestic production (such as Huweidi suggests), can increase costs owing to the need to stockpile and organise a wider range of spare parts; diversification remains a sensible policy, but need not exclude local industry, which can produce a number of those parts.

A plausible argument might be to adopt an export-oriented industrial strategy rather than an ISI one, in order to avoid the problems of massive import-substitution and to opt for high commercial returns instead. However, for exports to be competitive in world markets, they must not only be relatively cheaper but also comparable technologically. In fact, exports must enjoy even higher technological levels than some of the items produced for local consumption; even if the arms producer aims for the low- to medium-technology market, this will require substantial R&D effort in order both to keep pace with competitors and to reduce manufacturing costs (and product prices). Furthermore, the rapid pace of technological change builds a cycle of obsolescence into arms production, which means that the producer will eventually need either to maintain constant R&D or recommence imports on a significant scale. One way to save on R&D yet maintain technological quality might be to set up local subsidiaries of foreign defence (or industrial) companies or to undertake more subcontracting for the latter – this is what NICs such as South Korea and Singapore have done in effect.

There may be a number of objections to too close an industrial and economic relationship with the AICs in the field of arms production in the case of the Arab countries. Principally, the primacy of the national security incentive requires greater independence and restricted foreign access to sensitive sectors (though the number of foreign experts assisting Arab armies undermines this argument). It also indicates that total military spending and hardware requirements will remain high, meaning that the foremost task of the domestic defence industry will be to supply local needs, and so it has to enjoy wider scope and greater capabilities than those normally provided by subcontracting or subsidiary arrangements. Nonetheless, developing indigenous production capability might not necessarilly mean reducing technological dependence on external sources, specifically the AICs. Thus an official analysis of the Indian defence industry in 1979 conluded that despite many years of experience 'the implications are that, for modern and sophisticated weaponry, dependence upon the four major arms producing nations (USA, USSR, France, Great Britain) cannot be avoided'.[21] This conclusion, of course, has implications for Arab military industrial strategy.

The reference above to the NICs has special significance. To begin with, it is they which have been most successful with military exports. This is owing in part to the structure and organisation of their defence industries, and to their R&D effort (in terms of investment and organisation). But it is largely owing to the fact that they have well-developed civilian industries, which enable them to 'feed' the defence sector with primary materials and achieve import-substitution in various categories, and so to obtain the highest returns for their local economies (as well as balance debt with growth). Thus it is no coincidence that the NICs have the highest ratio (by value) of local defence production to arms imports.[22]

Furthermore, one advantage leads to another: by enjoying a sound industrial base and steady economic growth, most NICs are in a position to impose commercial arrangements with large, binding offset clauses. They are in a position to do so thanks to the volume of their production, and so gain credibility, but more importantly are in a position to benefit from offset terms. South Korea, for example, has been able to impose sizeable military and civilian counter-purchases on the USA, even though it relies on the latter for licences, technology, and parts and often operates as a sub-contractor. Indeed, its offset deals with individual American companies have been more rewarding than general government-to-government commitments.

The foregoing is also demonstrated in the case of Israel, which has the sixth highest offset percentage in the world, worth $1,384m out of $6,038m in arms purchases:[23] the official American commitment made in 1981 to buy back Israeli goods worth $300m annually was not implemented, but a growing volume of company deals pushed total Israeli sales to the USA to $400m in 1988.[24] Treating Israel as a non-NATO ally opened American markets to it, of course, but Egypt has enjoyed neither comparable sales nor high offset levels, despite being made a non-NATO ally too. In its case, as with Saudi Arabia, which has been unable to extract much benefit from its offset agreement with Britain or to attract substantial company investment, the weakness of the domestic industry or economy is at fault.

The Debate in Practice

Paradoxically, civilian–military linkage in countries like Singapore and Brazil is most important not in providing materials and components thereby allowing economies of scale in the 'feeder' industry, but rather in offering an organisational and structural example, managerial and marketing skills, and economic thinking. Thus they have been able to set up their defence industries as distinct economic sectors requiring minimal investment, requiring less institutionalised civilian–military industrial exchange. This helps to explain why the Brazilian aircraft company Embraer can operate with 4,800 employees (or the AFV-maker Engesa with 4,500), in contrast to the 40,000 at Hindustan Aircraft Industries.[25]

That said, the advantages of having a sizeable civilian industry and extensive military R&D effort do not accrue only to export-oriented producers. India, which follows ISI strongly, was importing only 30 per cent of the parts for the *MiG*-21 and 3–5 per cent for its *Vijayanta* tank by the mid-1980s, and is scheduled to reach import levels of 25 per cent for the *Milan* ATGM and 5 per cent (from 90 per cent initially) for the T-72.[26] This has resulted from its prolonged investment in R&D, and to the general level of its industrial base, which have enabled it to raise the local content constantly.

Possibly as important to India's success has been the fact that it can ensure long production runs, since it is supplying its own, large armed forces. By 1984 some 63 per cent of the IAF's aircraft were locally-produced, as were 65 per cent of army tanks.[27] In Brazil, 77 per cent of the air fleet (including helicopters) and virtually all wheeled armoured vehicles, light tanks, rockets, missiles, and small naval vessels were of local origin; moreover, all planned future air force procurements were of Brazilian-made aircraft.

Thanks to its industrial efforts, Brazil was exporting military *matériel* worth 21 times more than its arms imports by 1982. This may have been possible in the case of Brazil, which has a low external threat profile, but not for India, which remains in a state of tension with Pakistan and plays an increasingly visible role as a regional power. Pakistan is compelled to maintain far higher levels of defence spending and of arms imports, which transcends the output ability of its domestic military industry. Thus it is increasingly seeking export markets in order to boost the economic feasibility of its defence production. This is also connected to its need to invest even more in R&D and overcome shortcomings in its industrial base, for which it needs to improve the finances and raise the general profitability of its defence industry, and to increase co-operation with foreign companies and governments.

Issues of feasibility naturally impinge heavily on industrial strategy. As AOI head Orabi notes, the aerospace industry, for one, must have a production run of at least 120 aircraft to be successful.[28] This has been achieved only in the case of the *Tucano* trainer, in which local input reached 73 per cent, but would not have been possible even then had it not been for the guaranteed sale of 80 aircraft to Iraq. The example of Turkey is particularly pertinent in this context, therefore, as it has attained this level of success single-handedly, so to speak: it has launched several ambitious arms production programmes, most notably the assembly of 160 F-16s and the production of some 1,700 AFVs. The contrast implied here operates in two ways. On the one hand, Egypt (or equivalent Arab states, like Iraq and Syria) is in itself a sufficiently large consumer to justify large production runs – indeed, it plans eventually to acquire a total of 160 F-16s, in four batches. But it simply cannot afford to do so for financial reasons: it cannot raise the funds necessary to launch a run of 160 aircraft in one step, but neither can it afford the greater unit cost of running separate, 40-aircraft batches. Thus attaining feasibility would tend to devolve on to a richer country, such as Iraq (pre-1990) or Saudi Arabia, or demand joint Arab acquisitions.

The preceding review leads to a number of preliminary conclusions, which must be factored into any Arab strategy for military industrialisation. Most importantly, it is obvious that there is no single solution, no simple choice between opposing industrial and import policies. That is, there cannot be a single approach to import-substitution or export-orientation, foreign involvement (in R&D, technology transfer, or manufacturing), or technological level of products (low/medium/high), that is applied equally at all levels of the industrial process and in all its branches. Rather, any strategy must be differentiated, involving varying methods and policies in distinct areas of military industrialisation. This is particularly so if indigenous defence production is meant to respond to national security needs, or to implement an ISI approach. This differentiated approach moreover lends itself to any inter-Arab co-operation in arms production or industrial development, as there could be a division of functions or specialisations between several countries. This would not only make better use of existing resources, according to their geographical distribution and ensure wider benefits, but would also avoid the emergence of excessively large industries in the smaller states.[29]

The fact that import-substitution is regarded as being generally effective in reducing arms imports in practice commends such a policy. However, as Looney notes, success in replacing one import commodity might only come at the expense of retarding R&D and production of other items if the local industrial base is weak.[30] Thus much depends on widening the technical basis for local production, even if this initially involves greater foreign participation in some branches. In this 'vertical' growth, different sectors can be developed at varying paces, employing distinct policies, as long as they are integrated within a total strategy. Breaking down production into several elementary (or not so elementary) processes moreover suits the Arab countries (and most LDCs) because the manufacturing techniques tend to be more labour-intensive at that level.

The benefits of a differentiated approach become especially apparent in comparison to the seven stages of military industrialisation enumerated by Looney (discussed in Chapter 1); (1) local servicing and maintenance for imports; (2) licensed production with substantial outside assistance; (3) local assembly of imported subassemblies; (4) local assembly of imported components (to make subassemblies) and possible re-export; (5) local manufacture of components using imported materials; (6) local production of raw materials; and (7) complete indigenous production (also including design).[31] The first three are especially dependent on continued imports; the last three less so. In several arms-producing LDCs, the local defence industry operates at several different stages at the same time. This variation could moreover operate, and indeed often does, within any particular branch of the industry, and involve the civilian sector in varying degrees in parallel.

A question that arises, therefore, is where to place the emphasis, and which categories offer the most effective point of entry. An Egyptian defence industry official, for example, suggests that aircraft present the ideal starting point because they 'combine 95 per cent of existing high technology in the world. When we speak of an aircraft we are in fact speaking of everything we can imagine of world technology, whether in raw materials, processing, electronics, manufacturing technologies, or production systems . . . if we start with this industry, we will absorb the greater part of all technologies'.[32] The truth of this statement cannot obscure the fact, though, that the choice is not a subjective one – it is objective capabilities and constraints that decide the issue. And these relate principally, once more, to availability of funds, level of the general scientific and industrial base, and extent of the local R&D effort.

In summary, the ISI approach may be broadly suitable for Arab needs, but not in its simplified form. Rather, it needs great care in defining goals and setting criteria for feasibility and effectiveness. Thus equipped, an ISI strategy should assure economic benefits that balance the increased technological dependence that Huweidi correctly sees as resulting from indigenous defence production. However, a primary condition for industrial and economic feasibility, especially in the case of Egypt as matters stand, is to expand the base of production and extend runs, without which ISI is far less efficient.[33] Conversely, the closer any Arab arms producer comes to developing the local defence industry on a country basis, the more likely it is to organise itself along commercial lines and seek foreign markets and profits. With spiralling costs of military technology this becomes a necessity. But in all cases, inter-Arab co-operation would either make ISI strategies more realisable, or provide a primary target market for an export-oriented approach.

Conclusion

It remains to be stressed that an essential factor in the long-term feasibility of indigenous defence production, and indeed of general economic development and national security, is the social and political context within which it takes place.

It would be too much to claim that arms production cannot develop beyond a certain point in the Arab countries without far greater social and political freedom; if there is enough determination and professionalism any country with a certain minimal level of funds and competent manpower can achieve notable results (witness Iraq prior to 1990). Nonetheless, a largely independent technological and production capability cannot be attained, especially in the Third World, without high levels of mobilisation, participation, and motivation at both the collective (pan-Arab) and the individual country level. To take a concrete example: increasing the role of universities in R&D is viewed as a major task of Third World arms producers, especially in the light of their substantial investment in education, but academic research is particularly dependent on guaranteed personal freedoms and on free access to sources and information.

Thus participants in industrialisation – at all its levels and in all its branches – must perceive real stakes, and they must have the opportunity to develop skills and knowledge. Such drive is most obvious in Egypt, which has both the heritage and the incentive for extensive industrial (including scientific) development but lacks the funds. It is progressively weaker in Iraq (up to 1990) and Saudi Arabia, though, owing both to the discouraging impact of their social and political systems and to the undermining effect of their welfare systems. (This is not to criticise the welfare role of the state as such, but to dispute its mechanisms and consequences in the Arab case.) Any of these states may perpetuate its current policies, nonetheless, but it will be compelled to confront the issue increasingly as it seeks to attain greater commercial feasibility. It will also face the challenge as it expands inter-Arab co-operation, if only because the economics of its local defence production compel it into regional integration.

Mary Kaldor sums up the issue most succinctly: 'Technology cannot be treated in isolation . . . It is the product of a particular set of institutions nurtured in a given political and economic environment. Technical change is also social change.'[34] There is no single institutional model or socio-cultural and political norm to follow, and the Arab countries may develop a variety of their own indigenous responses to this need, but respond to it they must if they hope to achieve lasting industrialisation or development.

12

Towards an Alternative Strategy for Arab Military Industrialisation

Before the Gulf crisis and war of 1990–91, three out of 21 Arab states actually possessed, or were developing, significant defence industries. These were Egypt, Iraq, and Saudi Arabia. Of the three, Egypt has the longest-established industry, with a history of ambitious efforts at local design and development in such major areas as aircraft and ballistic missiles. Iraq was a newcomer, but under the pressure of combat needs during the first Gulf war (with Iran) it made rapid strides that brought it almost on a par with Egypt in terms of variety (except aircraft), volume, and local content. In contrast, Saudi Arabia had only modest productive capabilities, but possessed the financial means to implement the broad plans it had underway for military industrialisation.

Of course, the aftermath of the Iraqi invasion of Kuwait in August 1990 may yet cast a wholly new light on Arab military industrialisation efforts. The Iraqi arms industry was extensively damaged during the Gulf war and will suffer a shortage of funds as the country pays punitive reparations in the years ahead. Sanctions imposed by the UN and others will also constrict the flow of components, tools, and military technology to the sector, in addition to the limitations that might be set on the size of Iraqi armed forces and on the types of weapon they are allowed to field.[1]

Nonetheless, a number of compelling factors may lead the main Arab arms producers to renew their efforts at military industrialisation. Although the restrictions imposed on Iraq are both severe and sweeping, they may not entirely impede reconstruction of at least part of the local arms industry. Indeed, the embargo may spur an even greater emphasis on indigenous production in order to raise autonomy, however slightly. Certain sectors might prove easier to refurbish and resupply: those that involve low-tech manufacturing processes or products (ammunition and certain spare parts, for example), or that overlap with civilian industry (such as electronics).

Furthermore, a number of major arms producers have a vested interest in maintaining at least part of the Iraqi market, whoever is in power; private companies in the West or sympathetic third countries might also assist in the transfer of military technology to Iraq. Former Soviet republics in particular are likely to be sensitive to the defence needs of such Arab states as Iraq and Syria, partly for political reasons, while a range of countries from China and North Korea to Brazil, South Africa, and the East Europeans need export markets.

While national security would be the driving force behind any future Iraqi arms production, the main incentive for Egypt is to preserve the large investment it has already made in military industrialisation. In practical terms, commercial and economic considerations, in that order, would shape Egyptian policy. In this context, the opportunity for Egypt to offer itself as a protector of Gulf security and to project itself as an Arab regional leader once more (by pursuing the Palestinian-Israeli peace process and reviving the disunited League of Arab States) would provide the justification for continued arms production and a renewed export drive. The fact that Arab ground forces – in the Gulf and elsewhere – will still need low-to-medium technology military hardware leaves some scope for Egyptian industry; indeed, this requirement is also an argument in favour of Iraqi and Saudi industrialisation.

Components of a Strategy

That said, there are strong reasons to alter the main thrust of local military industrialisation. The main effort might not be to produce as wide a range of hardware as possible, but to achieve the greatest financial savings and the most efficient economic (civilian) utilisation of military expenditure. Implied here, for instance, is that although competition against imports (or achieving exports) might be beneficial, import-substitution as such is not as important as using financial resources most effectively, in a way that ploughs back as much as possible into the local economy and development.

In other words, the task is to identify certain categories of military expenditure which can be replicated indigenously more easily and that will involve greater locally-added value and less import content. Providing services rather than hardware is a prime example, which, moreover, has the advantage of being labour-intensive, and thus suited to Arab economies, and also focused on the transfer of human skills and manpower development at the same time.

Leaving general aims aside, a number of key operational objectives must be defined within the realm of defence production. One of these is to generate funding for the industry, whether from the state, private sector, or external sources. Naturally the source and volume of funding are affected by the scope and scale of the industry, and by its orientation (towards exports or import-substitution) and technological level. A second objective is to acquire technology, again through a variety of possible means, such as foreign transfer or local R&D. Within this general term, though, it must be defined whether the specific purpose is to support particular local production programmes, develop an indigenous modification or design capability for application to the types of hardware used or produced locally, or generally keep pace with world military technology in all major fields (such as aerospace). Of course, each purpose will relate to a different industrial strategy and involve varying levels of funding. A third operational objective is to develop the scientific, technical, and industrial base in order to support defence production (and R&D). This, however, is a very broad aim, and so a more concise, relevant target is to develop the requisite human skills and manpower for an ambitious defence industry.

Military Needs

This is a difficult point to decide, as the needs of most modern armies are likely to be defined automatically in broadly similar terms. From advanced interceptor aircraft to night-vision goggles, from tanks equipped with full electronics suites (FCS and laser range-finders), SA missiles and radars, to precision-guided munitions, these and many more items, from the pedestrian but no less essential combat-support equipment (bulldozers, trucks, and protective clothing) to sophisticated training simulators, have come to form a basic part of many Third World shopping lists.

In addition to actual need (or real ability to absorb and utilise), a major influence motivating such acquisitions is the desire to enhance domestic prestige and regional status. And both psychological and strategic factors underlie the impetus to match armament developments in neighbouring states, thus fuelling a regional arms race that may actually be beyond local economic and technical means or military requirements.

Arab military needs must be defined nonetheless. They must be defined critically, moreover. A powerful strand in Arab military thinking is that local capabilities must be as up-to-date as possible in comparison with international trends, and that Arab armies and industries should strive to acquire and absorb the most modern equipment and technologies available. This view is powerfully reinforced by the fact that regional neighbours (such as Israel) or out-of-area forces (American forces in the Mediterranean and the Gulf), possess state-of-the-art weapon systems. The net thrust is to suggest that Arab armed forces must have comparable capabilities in the same categories of conventional arms. (Interestingly, the same logic is not applied to non-conventional weapons, in which the Arabs have consciously developed chemical weapons and ballistic missiles as an alternative, countervailing force to the Israeli nuclear capability.)

Evidently, the ultimate result is to encourage high capital investment and rapid hardware turnover (in order to keep pace with external developments), and also to obscure certain areas of need that require added attention. Indeed, the Arabs may be overlooking alternative methods of improving military effectiveness and disregarding their own strong points.

This is brought out clearly in a number of ways. As shown in Chapter 3, the ratio of equipment to manpower in some Arab armies is excessively high, pointing to low utilisation and leading ultimately to high attrition and poor performance. Conversely, changes in combat doctrine may affect force structure far more, and produce greater benefits, than increased military spending (on equipment).[2] This was undeniably the Egyptian experience in 1973, when infantrymen using simple RPG-7 anti-tank launchers and *Sagger* ATGMs (as well as ZSU-23-4 AA guns, man-portable SAM-7s and ageing SAM-2/3s) repulsed massive Israeli armour counterattacks. The last example – as well as the limited Syrian–Israeli 'rehearsal' in 1982 and the far more massive confrontation in the Gulf war of 1991 – also highlights the potential future for combat doctrines relying heavily on PGMs and associated high-technology systems.

State-of-the-art munitions and support equipment involve much high-tech and financial investment, it is true, but even so remain less costly to develop and produce than major weapon systems (such as advanced fighter aircraft) and their associated design and manufacturing technologies. PGMs and high-tech 'mini-systems' are moreover relatively easier to absorb and use, and therefore cost less time and money in training, especially with the use of simulators, for Third World personnel. Indeed, the whole issue of training is another main area for major potential improvement at relatively low cost: it is estimated that up to 50 per cent of bad equipment performance is due to operator error.[3]

At a broader level, the fact is that the proficiency of any country's armed forces is related to the nature of its manpower and economic base, and cannot be far in advance of the general level of local education and scientific/technical culture. This is not to say that the national army cannot be highly efficient and effective militarily: the Vietnamese fought American forces to a standstill, but this was because of efficient mobilisation of available resources and formulation of strategies and tactics most suitable to existing capabilities and conditions, rather than to advanced armament.

In the Arab case, even if the armed forces are seen as the vanguard and cutting edge of modernisation (in terms of scientific and technical thinking), they cannot escape completely the reality of a 42 per cent illiteracy rate (in 1985), for example.[4] After all, personnel have to come from a given population base, and military development (both quantitative and qualitative) depends on local generation of a number of needed resources, thus setting an inbuilt limitation to the type of armament and warfare adopted by Arab armies. This relationship has become even more binding with the development of military technology. That the Arabs have nonetheless achieved intensified military development is indisputable, but it has come with an even higher price and increased external dependence.

The implication here is, therefore, that the Arab states need to reassess their military needs more carefully and realistically. This will lead both to improved combat capabilities – as existing resources are used more fully – and to financial savings as waste is reduced and unnecessary items of expenditure are avoided. Much more depends on appropriate doctrine and strategy, and on the 'human' dimensions (organisation, command, and training), than on hardware. This was the painful lesson of the 1967 Arab–Israeli war; one that has been applied with increasing degrees of success in 1973, in the battles between Syrian commandos and Israeli armour in 1982, and in Iraq's defence against Iran after 1982. Ultimately, the whole exercise hinges on the impact of leadership perceptions and strategic thinking: these reached an epitome with the manner in which the Iraqi President Saddam Hussein decided on the invasion of Kuwait in August 1990 and then 'managed' the subsequent crisis and war.

Types of Product and Technology

Critical as the role of human dimensions may be, part of the Arab response should also be to develop certain areas of military hardware and technology. The main question here is what types of defence product the Arab states should produce, and how far they should go in acquiring an indigenous capability to design and develop them. Aircraft and armoured fighting vehicles pose the greatest challenge, though all high-tech systems present a problem. Conversely, low- to medium-technology hardware, such as infantry weapons, rocket artillery, and conventional munitions, are evidently less problematic, and are already in production (and even development) in a number of Arab and other Third World arms-producing countries.

In a sense, the answer to the above question is already provided in part: the Arab states can, and should, produce a large proportion of their 'combat consumables' locally. They should, moreover, embark on large-scale production of a large variety of non-combat support equipment, which at present is rarely manufactured (let alone designed) locally. But this broad category is essential to the functioning of any army, and accounts for a substantial share of expenditure. On average, non-combat equipment is cheaper to buy than weapon systems: it involves far less advanced technology, fewer components, and easier production techniques generally. Cheapness in turn encourages reliance on imports. However, the option of producing support equipment locally is sensible because it does not afford foreign suppliers such clear leverage over the recipient's political and strategic decision-making and its war-waging capability, in sharp contrast to the leverage they gain through supply of main weapon systems.

Nonetheless, the fact that non-combat support equipment is relatively easier, and cheaper, to produce and is required in large quantities should be taken as an incentive to establish a local Arab production capacity. This is made even more feasible through the degree of overlap that exists between the military and the civilian sector in this field – such as tractors, bulldozers, trucks, jeeps, engines, hoists, tooling, and batteries – which will allow longer production runs and spread the costs of R&D or of setting up new processing plants for raw materials or common components.

Assuming that the production of combat consumables and support equipment presents no particular problem, the question then narrows down to what degree of involvement should be undertaken in the aerospace missile, AFV, electronics, and other high-tech divisions of defence industry? Affecting the response here, on the one hand, is the view that high-tech weapon systems incorporate an overwhelming proportion of all modern design and process technologies, and so embarking on such fields as aircraft or AFV production would introduce Arab industry to the widest range of scientific and technical know-how.

On the other hand, high-tech weapon systems are also the most expensive to design and produce, for the same reason. This leads to growing lead times (between concept and end-product), during which subsidiary technologies and components may undergo more rapid change that may lead in turn, ironically, to early obsolescence. The latter phenomenon becomes especially apparent when improved sub-systems suddenly render ineffective weapons that have received substantial investment of time and money. A case in point is the British LAW-80 anti-tank infantry weapon, a development of which was suspended in late 1989 after £400m had been spent, because it had already become useless against Soviet tanks equipped with cheap, add-on, reactive armour plates. In general, the daunting economics of developing and producing an indigenous, modern, weapon system are best demonstrated by the fact that even the West European AICs have repeatedly had to form consortia to design new combat aircraft (*Jaguar*, *Tornado*, and *Rafale*), and that only the USA and the former USSR still produce all system components by themselves.

Against this background, the suggestion here is for a varied approach in choosing types of military product and technology for local absorption. Some systems might be produced and eventually designed locally, but others might best be imported or made under licence. The point is to delineate those types and areas in a way that takes due consideration of financial and technical requirements, Arab capabilities, world armament and trade developments, and the impact and significance of changing technologies.

But at the outset, it does not seem feasible for the Arab countries to develop a completely indigenous design and production capability for such major weapon systems as aircraft and AFVs. The financial investment is simply too massive and the risk of failure too high for countries seriously lacking in their industrial, scientific, and manpower base – especially as cyclical obsolescence will constantly threaten any progress made and pose new challenges. The experience of India's *Vijayanta* and LCA programmes – which have been dogged by delays, technical hitches, under-performance, and cost over-runs – shows the dangers and difficulties even when the country has invested continuously in its industry and R&D. Both projects have nonetheless been qualified successes, yet even so India has turned once more in recent years to acquiring foreign-made or designed aircraft and AFVs (Mig-29, *Jaguar*, and T-72). Somewhat similarly, the Brazilian–Italian joint combat aircraft has not achieved any commercial success so far, indicating an added risk of financial infeasibility for major weapon systems that are not state-of-the-art items produced *and used* by the AICs.

However, this is not to deny the advantages of producing major weapon systems locally. Indeed, it is useful to continue licensed assembly or production, given two conditions. The first is that programmes be structured to ensure maximum economic feasibility as a primary aim: the example of Turkey's F-16 programmes shows how local assembly can be made cheap and beneficial – principally through extended production runs but also by high participation by local industry. This will require greater inter-Arab co-ordination and standardisation. The second is that a well-planned R&D effort must run in parallel with production, both in order to assure technology transfer within the specific programmes actually underway, and to help to keep pace with world developments in the field. Thus even if production is wound down, the Arab defence industry could retain a potential for future activity in major weapon systems, by investing in an appropriate R&D base, though not in a large productive one. Such a capability would also allow Arab input into modifying as well as manufacturing sub-systems and components, in order eventually to produce 'hybrid' major weapons and upgraded.

The R&D factor suggests that a special effort might be placed in acquiring know-how in particular subsidiary areas that are related to the development and production of major weapon systems. These could include supporting technologies and applied sciences, such as special chemicals and metallurgy, or industrial techniques, such as casting large foundry pieces (tank turrets or hulls). Here the main effort should once again be focused on R&D, in order to keep abreast of world scientific knowledge and technical processes. However, it would also be useful to engage in actual development and production, *if the results could be usefully and economically used in local defence production and/or in the civilian sector*. As both sectors can benefit from scientific advances in primary and raw materials, there is evidently considerable room for military-civilian overlap in this area, which should be deliberately exploited.

At present, civilian industrial R&D is severely lacking and underfunded: but dedicated R&D in applied sciences and supporting technologies could be economically feasible and financially rewarding if it fed both sectors (civilian and military). Without such co-ordination, however, research into subordinate areas earmarked for specific weapon projects can be excessively costly. Furthermore, a narrow focus could not be sustained indefinitely unless (at best) it was feeding into an ongoing production programme, whereas a broad, diverse R&D base would provide both flexibility and continuity in subsidiary specialisations.

The issue of major weapons development and production highlights a second vital area for Arab concentration. This is the role of high-tech sub-systems in those weapons. It is components such as computers, electronic sensors and displays, and communications equipment that make aircraft or AFVs combat-effective and that act as force-multipliers in modern warfare. They are the heart of avionics suites, fire-control systems, and tactical command centres, and they can account for up to 60 per cent of the production cost of a modern major weapon. These components moreover incorporate a wide range of technologies, from general electronics, through infra-red, laser, and millimetre-wave guidance, to optronics. They can also apply to manufacturing techniques, as in the case of robotics, which has both design and process uses.

An immediate conclusion is that the Arab defence industry might focus its efforts on acquiring some of these technologies, and on developing indigenous design and production capabilities in them.

Yet it is precisely these fields that are the most difficult to replicate locally, as they are highly demanding of human skills and a scientific base, and their production is especially capital-intensive. They are, moreover, the target for much of international R&D, both military and civilian, as they are the cutting edge of modern technology and industry, and so any local high-tech effort will be particularly exposed to rapid change and obsolescence.

Nonetheless, if the Arab states are to make a substantial, long-term commitment to developing their indigenous defence industries, then high-tech sub-systems should be a primary target, as they are 'the wave of the future'. (For convenience, these will be referred to as the 'high-tech complex', which includes actual electronics hardware such as computers and communications equipment and associated sciences such as thermal imagery (IR) and fibre-optics.) In addition to earlier comments on their role in major weapon systems, the view that the Arabs should focus on high-tech systems is supported by other international and military trends.

On the one hand, an ever-growing proportion of global activity revolves around the use, or sale, of information technology, and of high-tech systems more generally. This is having sweeping impact not only on the patterns of world economy and trade, but also on their content, as information technology is expected to occupy a massive share of global GNP in the twenty-first century.[5] The trend also affects national administration, and will ultimately influence foreign policy and inter-state relations.[6] For similar reasons, the importance of 'artifical intelligence' in military applications has gone well beyond its use in improving the combat performance of specific weapons or of providing tactical C31 (command, control, communications, and intelligence). It is now seen as 'the tip of an iceberg representing a huge body of information technology which is becoming a pivotal element in many key areas of importance to international security, command and control, strategic defence and verification, to name but a few'.[7] The Arab states will fall behind in this field at their peril.

On the other hand, the high-tech complex has major relevance for current trends in military hardware.[8] Increasingly, combat consumables incorporate a 'smart' capability: the ability to sense and distinguish targets and to direct themselves towards the latter. This no longer applies only to guided missiles of all types, though even they are gaining greater 'fire-and-forget' capability. More importantly, it relates to unguided munitions such as free-fall bombs, artillery shells, and surface-to-surface rockets. The first are increasingly being equipped with add-on guidance kits to turn them into 'gliding' bombs and increase their stand-off range (raising aircraft safety) – as Iraq has done – while all types are being produced in cluster/

dispenser versions which release precision-guided sub-munitions, especially anti-tank ones.

Gun and rocket artillery are also moving in the 'smart' direction, using a variety of technologies (IR, millimetric, and laser) to fire operator-directed or fire-and-forget PGMs. Furthermore, conventional guided missiles are being designed with improved guidance technologies in order not only to improve their precision but to reduce their vulnerability to counter measures. The use of optical fibres in ATGMs is a case in point. Finally, regardless of the eventual fate of the American SDI programme, 'star wars' technology is leading to the next generation of military hardware, employing lasers, directed-particle beams, and kinetic-energy rail guns.

In other words, the Arab arms producers should learn from Egypt's experience in the 1950s and from the more recent lessons offered by the NICs, to grasp the trend of modern technology and anticipate its implications for defence and industry. It is only on this basis that long-term strategies should be set, and current policies formulated. There are strong arguments for making the major investment in R&D and a broadened scientific base (including manpower development) in order to keep pace with international trends. At the very least, it must be understood that even conventional combat consumables are coming to incorporate high-tech elements. That is, the Arab defence industry cannot even hope to continue production of low-tech goods indefinitely, assured of local demand, as these become obsolete and national armies refuse to acquire them. The influence of the high-tech complex is spreading to all levels of military and industrial activity, and the implications must be addressed squarely.

Maintenance, Parts, and Other Services

As stated earlier in this chapter, a principal aim for the Arab states should be to achieve financial savings and reduce total military expenditure in a manner that contributes to GDP. So far, much of the debate about military industrialisation, and most local defence industry activities, have focused specifically on the production of hardware and the use of associated technologies. Yet it is a fact that a major part of defence spending goes not to the purchase of equipment, but to provide the wide variety of associated and unassociated services: maintenance, repair, overhaul, training, spare parts, engineering design, and construction. Indeed, the more advanced the weaponry and modern the army, the higher is its tail-to-teeth ratio.

In this regard, the situation of Arab armies is contradictory. On the one hand, the principal Arab armed forces have amassed large numbers of sophisticated land, sea, and air weapon systems. Their equipment-to-manpower ratios are high, which would seem to indicate a high level of proficiency and combat effectiveness. However, these ratios are in fact well beyond the managerial, organisational, training, and support (logistic and repair) skills of the Arab states concerned. One result is to risk rapid equipment attrition and substantially reduced effectiveness in any combat situation, and more generally to impose a maintenance and operational burden on Arab armies that they cannot meet.

The attendant financial loss is considerable: due both to the wasting of the original investment in purchasing hardware, and to the fruitless expenditure on servicing subsequently. This is not to mention the national burden of servicing external debts incurred in importing the equipment in the first place. An alternative consequence is to rely heavily on foreign personnel, whether hired direct or under contract to foreign companies and governments, to provide these services and even to operate hardware. This is most obvious in the Gulf states, but applies in varying degrees to most of the other Arab countries too. Acquiring such assistance is costly, and seriously questionable in its impact on national security, as is the other alternative of shipping equipment back to its manufacturer for routine overhauls.

These observations are brought out more forcefully on referring to practical examples. To take one, Egyptian officials have expressed their disquiet at the high cost of follow-on support (maintenance and spares) for their 80 F-16s, which consumed $1bn of the $1.5bn disbursed for the programme in each of the two years up to 30 September 1988.[9] All of this amount came from American military assistance, as Egypt stated that it could fund only a portion of the support for American-supplied equipment through to FY1992. Summarising the problem at an earlier date, President Mubarak asked: 'How can we provide maintenance for equipment worth more than $30bn or $40bn, for instance, with a budget of only $300m?'[10] Indeed, the extent of the problem is even more staggering, since the cumulative maintenance cost throughout the operational life of equipment can reach *four times its original purchase price*.[11]

The nature of the problem emerges even more dramatically from a breakdown of Saudi military expenditure. According to one American-based report, of $50bn worth of Saudi military purchases from the USA in 1953–88, $3.6bn was spent on spare parts, $13bn on support (including training and advice), and a staggering $20.7bn on construction – leaving $12.7bn, or roughly 25 per cent, in hardware acquisition.[12] Since then, the Saudi-American deal for the sale of 315 M-1 tanks has been presented for Congressional approval: its $3bn total value includes $1.28bn for tanks and other vehicles (42.7 per cent of the total package), and $1.81bn in various forms of support and follow-on (57.3 per cent).[13] The latter figure includes

$904.9m for spares (30 per cent); $300m for ammunition (10 per cent); $142.9m for facilities design and construction (4.8 per cent); and $79.8m for training aids and devices (2.7 per cent). (Maintenance, training, and advisory services are not included, and doubtless will be accounted for in separate contracts.) Especially notable are the shares for spares and ammunition, which indicate the potential for replacement of imports by local production. Significant too is the share of training aids, which though small in total proportion is nonetheless appreciable and represents an area for development by Arab defence industries.

The emerging picture is that a massive slice of Arab military expenditure is going to the various forms of follow-on support for hardware. Incidentally, the same applies broadly to armies using Soviet or Eastern-bloc equipment, though contract details have never been released in nearly four decades of Arab-Soviet military exchange. It is plausible to assume, though, that Soviet services (like their equipment) are generally cheaper than their Western counterparts. Furthermore, Soviet-supplied Arab countries have generally been those that are more security-conscious and less open to foreign inspection – such as Iraq, Syria, and Algeria – encouraging the assumption that they have also put more effort into providing more of such services locally (especially construction). Conversely, Egyptian experience with the USSR in 1955–72 was negative in this respect, with general Soviet reluctance to transfer skills or permit local overhauls or parts production. (This attitude may change in the 1990s, with greater Soviet eagerness to win foreign markets.)

Thus there is a major potential for the Arab states to achieve massive savings by providing an increasing proportion of military services locally. To show how important this can be for war-waging capability, the Egyptian Minister of State for Defence Production General Sayyed has asserted that 90 per cent of tanks hit in combat and 80–90 per cent of all damaged armour could be repaired and brought back into service if spare parts were provided locally.[14] Indeed, AOI head General Orabi places provision of maintenance and spares ahead of producing complete weapon systems in the scale of importance.[15] In addition to the operational dimension, the financial rewards of providing services locally are enormous, as is evident from the figures already cited for Egyptian and Saudi expenditures. These amounts would be spent within the Arab countries instead, thus saving hard currency, reducing debt, and making a substantial contribution to local GDP.

There would be an added security advantage too, since nationals rather than foreigners (or expatriates) would undertake an increasing role in engineering design and construction of military facilities, for example. The same applies to training: sending servicemen abroad in large numbers exposes the recipient to long-term political and ideological influence,[16] which may even be reflected in covert external links within the national armed forces of the Arab countries, with adverse consequences for the stability of government. There can be little doubt, in summary, that for states eager to assert their political and strategic independence, the elimination of after-sales dependence may be a more vital link even than ensuring arms supplies; after all, hardware is usually freely available in a competitive world market, but not all producers can provide effective services.

In final confirmation of the importance and scale of military services, the following figures on aircraft maintenance indicate the manpower levels (and hence the expenditure) involved. Using available data, the maintenance man hours per sortie for some aircraft in Arab service are: F-4E *Phantom*: 38; F-15 *Eagle*: 33.6; and MiG-21: 23 (per flight hour).[17] To keep a squadron of 18 combat aircraft in the air requires anything from 250 to 500 ground support personnel, as Neuman has shown in her study of the Iranian Air Force in 1978. In all, Iran had some 10,000 managers and technicians, 'a quarter of which had to have skills necessitating several years of training and experience', for base- and intermediate-level maintenance only (excluding main depot work) for 484 aircraft.[18] (This is not to mention the immense variety of spares that must be kept in stock.) Furthermore, considering the fact that the mean flight hours between failure is only 20–30 minutes for the above-mentioned types, and at best 72 hours for the E-3A AWACS, indicates just how crucial servicing is for continued operation. Unless such support can be provided locally, then in the absence of foreign personnel a recipient would only be able to operate its fleet for a matter of a few days.[19] Just how far reliance on foreign support can reach was shown by Iran, which had 25,000 American military-related personnel by 1979.[20]

For the Arab states, a principal advantage of concentrating on the development of services is that they are more labour-intensive than arms production. Furthermore, they involve the training of manpower and the production of skilled technicians and managers – services are one area in which automated machinery and robotics cannot replace human beings. The training aspect complements the aims of social and economic development admirably, since the key to the latter is human development, and could therefore be the result of a deliberate manpower strategy for both the civilian and the military sector. This is especially so as many of the maintenance and construction skills are readily adapted to either application. Indeed, this is the purpose of Saudi Arabia's aircraft and engine service centres, and the example of Jordan's own air force and national airline. By adopting a combined approach, moreover, the Arab states involved could utilise the cost of training more effectively, by broadening its base. In all, this might, well be the most feasible area for import-substitution and for utilising military expenditure in a way that achieved the greatest addition to GDP.

Conclusion

Many themes and issues have been presented in this study of the Arab military industry; too many, in fact, and with relationships that are too complex, to be reviewed or summarised here. Instead, this concluding section will reiterate the view adopted in the book, namely that the fundamental questions that must be dealt with by Arab decision-makers and planners relate to the connections between military industrialisation and economic development.

Specifically, the core argument made throughout and repeated here consists of three main elements. The first is that despite frequent references by officials or academics to the potential positive impact of local defence production on the economy and development, the effect has been negligible (and even detrimental on occasion) in the Arab case. Indeed, verbal or theoretical assumptions have rarely been fully thought out, and the purported military-civilian linkages have not been planned or actively sought, although they cannot be achieved without a conscious mental and physical effort.

Although the picture varies from branch to branch of the arms industry and from country to country, Arab reality is the following: the balance between arms imports and exports (and imports and exports of all types) has not been appreciably altered, and so hard currency has not been saved nor has the balance of payments or of trade improved: there have been few spin-offs for R&D and the training of skilled manpower, with the latter probably being diverted to the defence industry at the expense of the civilian sector; and the general lack of local R&D has undermined technology transfer generally and limited local content and the increase in GDP.

The second element of the argument is that, far from acting as a motive force by stimulating civilian industry and generating economic growth, the Arab defence industry has been seriously handicapped by the underdevelopment of the civilian sector. The problems of socio-economic development at the national level have actively slowed military industrialisation. In addition to the above-mentioned consequences, the defence sector has been starved of competent technicians and managers. The weakness of civilian feeder industries (as well as low R&D effort) has placed a ceiling on the local share of added-value and so raised costs and reduced commercial viability, in turn causing a shortage of funds (both public and private) and generating fewer returns for the economy.

In response, thirdly, it is the conclusion of this study that Arab decision-makers and planners must focus on three specific dimensions of industrial activity in particular in order to achieve the most effective military industrialisation possible. (To recapitulate, effectiveness in this context is measured in terms of local ability to provide basic defence needs, not only of combat hardware, but also of support equipment and technical services, in a manner that is not only the most rewarding financially but that actively contributes to development.)

The first dimension of industrial activity is the development of technical manpower, which is the key to development in all fields. Militarily this means a focus on the services side of defence requirements, both because it represents a major slice of spending and so can cause greater savings, and because the human element is the real key to improved combat performance, not hardware as such. Economically, manpower development can make the biggest contribution to GDP as well, with the civilian industry in particular benefiting from the training of technicians within the defence sector.

Another dimension of industrial activity is the conscious establishment or expansion of the key civilian feeder industries, which can both raise the local content of defence production and increase the economic feasibility of civilian production (by extending production runs or setting up larger facilities, and so reducing unit costs). This should be the point of entry and main arena for any ISI strategy, which moreover has the potential to bring about hard currency savings and expand GDP. Finally, there must be a far greater concentration of human and material resources in R&D and in select high-tech branches, as part of a broader effort to generate an industrial culture that must inevitably go hand-in-hand with a more liberal official attitude to scientific, educational, social, and political freedom.

Indeed, the last statement holds the ultimate key to understanding the processes and requirements of military industrialisation in a developing country: although the defence industry is a highly specialised sector in which it is occasionally possible to achieve a 'tactical' breakthrough by concentrating massive financial resources (when available) and willpower, there will nonetheless always remain an overriding need for a competent industrial, scientific, and human base at the national level that is able to provide long-term support, utilise the benefits and spin-offs of defence production, and generally provide an appropriate environment.

It is on this basis, and with the added benefit of a complementary approach that combines Arab resources as effectively as possible, that the Arab countries can attain greater independence individually and occupy an appropriate collective position internationally. It is potentially theirs, but in order to reach it they must initiate a number of fundamental processes that affect their economies, societies, and political systems. Seen in this way, there are no historical shortcuts either to military industrialisation or to development as a whole.

13

Epilogue

The Gulf war of 1991 raised the prospects of fundamental change in the patterns of conflict and military competition that have prevailed in the Middle East for decades. Above all, it appeared to demonstrate the futility of massive financial and human investment by Arab governments in armies and weapons inventories that could so easily be defeated by technologically and logistically superior Western forces. As importantly, the Gulf crisis brought an unprecedented degree of international co-operation, making the imposition of supplier controls on the transfer of military technology to the region feasible for the first time since the short-lived Tripartite Declaration of 1950. With the added loss of the USSR and the Soviet bloc as a strategic pillar of support and a counterweight to the USA and its allies (including Israel), Arab options for pursuing independent military development or waging armed conflict seemed limited in spring 1991.

Should these prospects be borne out, then the incentives and opportunities to establish and develop local defence industries in the Arab countries will diminish. Yet there are a number of reasons why the patterns of military development and weapons acquisitions might not change substantially and why arms controls might not work, suggesting that military industrialisation may remain an attractive option. In fact, acquiring or expanding an indigenous defence production capability might prove desirable for Arab governments even if arms controls and regional security are attained.

A fundamental problem with most arms controls thinking on the Middle East to date, including the initiative proposed by President Bush at the end of May 1991, is that it effectively overlooks the political root causes of conflict and instability.[1] A variety of motive forces have propelled Arab military development in the past, and until they are resolved or altered the broad patterns will not change sufficiently. The Arab-Israeli conflict (including the Palestine problem) and the Gulf-Iran balance are important areas of strategic concern, as are regional arms races and Arab inter-state rivalries. The re-introduction of Turkey into the Middle East balance and the proposed restructuring of NATO with a focus on the southern flank (Turkey) and out-of-area operations poses another potential challenge, as does looming competition within the Middle East over water and other natural resources.

The policies of global powers, especially of the USA, towards the region and its problems are a further factor: disregard of Palestinian rights, ambivalence towards Israeli occupation of Arab land, tacit acceptance of Israeli NBC and ballistic missile programmes, and support for Israeli military preponderance are foremost examples.[2] Yet such attitudes prolong tensions and help to destabilise the region, and may encourage the disadvantaged states to attempt to counteract them by pursuing rival military development – as may occur with Iran in the Gulf or Syria in the eastern Mediterranean.

At another level, militarisation of the Middle East has been a consequence of the way in which local ruling elites define goals and priorities. The experience of Iraq's involvement in two immensely costly wars in the space of a single decade is a case in point. Geopolitics – the search for a role and for local allies – has encouraged the emphasis on military security and defence rather than on regional co-operation in economic development and other fields, and still dominates the foreign policies of key Arab states.[3] The pattern of massive military expenditure has also been closely linked to the vested interests of many parties, both suppliers and recipients, in promoting the arms trade to ensure continuing profits and commissions.

The opportunity to achieve a fundamental change in these patterns offered itself in the wake of the second Gulf war, it is true, but is unlikely to be usefully exploited in the absence of a comprehensive approach to the region's many problems. A structured approach, such as that embodied in a Conference for Security and Co-operation in the Mediterranean and Middle East (CSCMME – along the lines of the European CSCE), might be appropriate, but runs counter to the narrow interests and selfish policies of most of the actors involved: Arab, Israeli, and Western. Already, the approval of several major new arms deals with the USA's allies in the region has undermined the basis for arms controls.

Thus the Middle East is more than likely to witness continued military expenditures at high levels, though possibly not as high as in the late 1970s and early 1980s. An added factor is the interest shown by numerous Third World countries soon after the end of the second Gulf war in acquiring high-tech weapons, munitions, and supporting equipment of the sort that performed so well against Iraq.[4]

In the event of continuing investment in military infrastructure and acquisitions, the Arab countries will continue to have an interest in developing indigenous productive capabilities. Indeed, even if total expenditure and acquisitions decline – under the impact of regional arms controls, financial constraints or political changes – the same countries might consider that they can make significant savings by embarking on certain areas of local military industrialisation. Whatever the course taken, Arab armies will always need a wide range of low- to medium-technology weapons and munitions, non-combat equipment, electronics-based C31 equipment, and support services. Not only can much of these be provided locally, but to do so would channel part of the defence spending back into local GDP and promote overlap with the civilian sector in ways most conducive to economic development.

Chapter Notes

Chapter 1 Introduction: Incentives and Determinants of Third World Defence Production

1 A solitary exception is the series of articles by Nabeel Ibrahim, 'As-Sina'ah al-Harbiyyah al-Arabiyyah: Nathrah Mustaqbaliyyah' ('Arab Defence Industries: A Futuristic Look'). *Al-Baheth al-Arabi*. London. Nos. 12–14, 1987 and 1988.
2 Saadet Deger, *Military Expenditure in Third World Countries*, London: Routledge & Kegan Paul, 1986, p. 152.
3 Ibid., p. 152.
4 H. Wulf, 'Developing Countries', in N. Ball and M. Leitenberg, *The Structure of the Defence Industry: An International Survey*, London: Croom Helm, 1983, p. 310.
5 Nabeel Ahmad, p. 87; and R. Vayrynen, 'La lente construction de l'industrie militaire egyptienne', *Le monde diplomatique*, October 1980.
6 Ali Qasim al-Oqabi, *Ath-Thawrah* (Baghdad), 11 October 1987.
7 A. Levite and A. Platias, *Evaluating Small States' Dependence on Arms Imports: An Alternative Perspective*, Cornell University, Peace Studies Program, Occasional Paper No. 16, p. 37.
8 R. Looney, *Third World Military Expenditure and Arms Production*, London: Macmillan, 1988, p. 177.
9 M. Brzoska and Ohlson, 'Arms production in the Third World: an overview', in M. Brzoska and T. Ohlson (eds.), *Arms Production in the Third World*, London: Taylor & Francis, 1986, p. 27.
10 Deger, op. cit., p. 153.
11 Looney, op. cit., p. 177.
12 Al-Oqabi, op. cit.
13 R. Vayrynen, 'The Arab Organization of Industrialization: A Case Study in the Multinational Production of Arms', *Current Research on Peace and Violence*, 1979, p. 66. Geoffrey Kemp is cited in C. Catrina, *Arms Transfers and Dependence*, London: Taylor & Francis, 1988, p. 233.
14 'Hal at-Tasni' al-Harbi Wasilah li Tahrir al-Qarar al-Siyasi?' ('Is Military Industrialisation a Means to Liberate Political Decision-Making?'), *Al-Arabi*, No. 325, December 1985. This is an option also discussed by Levite and Athanassios, op. cit., p. 46.
15 *Militarisation and Security in the Middle East: Its Impact on*

Development and Democracy, London: Pinter, 1989, pp. 71–2.

16 S. Neuman, *Military Assistance in Recent Wars: The Dominance of the Superpowers*, Washington DC: Praeger (for CSIS), Washington Papers, No. 122, 1986, p. 69.

17 Levite and Athanassios, op. cit., pp. 46, 48.

18 Deger, op. cit., pp. 175, 177.

19 Deger, op. cit., pp. 147, 158.

20 Op. cit., p. 28.

21 D. Whynes, *The Economics of Third World Military Expenditure*, London: Macmillan, 1979, p. 49.

22 As Wulf argues, op. cit., p. 325.

23 Op. cit., p. 155.

24 Op. cit., p. 312.

25 Ibid., p. 324.

26 Catrina, op. cit., p. 191.

Chapter 2 Arab Arms Imports: Strategic Context and Patterns

1 Based on *SIPRI Yearbook 1989: World Armaments and Disarmament*, Oxford: Oxford University Press, 1989, Appendix 6A, Table 6A.1, pp. 226–7.

2 M. Brzoska and T. Ohlson, *Arms Transfers to the Third World 1971–85*, Oxford: Oxford University Press, 1988. Table 2.1. p. 16, Table 2.2, p. 22.

3 Text of official statement of heads of state meeting in Damascus, *al-Hayat* (London), 7 March 1991.

4 Some of these issues are discussed in Yezid Sayigh, 'Arab Regional Security: between Mechanics and Politics', *RUSI Journal*, Summer 1991.

5 These policy aims are among four explicitly stated by the then Foreign Minister of Egypt, Butrous Ghali, in an interview, *al-Hayat*, 6 January 1991, and later reiterated by the new Secretary-General of the League of Arab States, Ismat Abdul-Majid, in an interview in *al-Hayat*, 16 April 1991.

6 It needs to be noted that these figures are not portrayed as authoritative. Many credible sources exist and have been referred to in this study, which are listed below. The main intention here is simply to take a consistent set of estimated figures that can be used to indicate patterns. Errors or variations between sources in the region of ±5 or even ±10 per cent will not change the picture (greater errors are unlikely in the literature). *The Military Balance*, London: IISS; *SIPRI Yearbook*, London and Oxford: SIPRI; *The Middle East Military Balance*, Tel Aviv: Jaffa Center for Strategic Studies (since 1983); Kasim Ja'far, *Mizan al-Qiwa al-'Askariyy fi Mintaqat ash-Sharq al-Awsat, 1984–1985 (The Balance of Military Forces in the Middle East Region, 1984–1985)*, Beirut: Arab Institute for Research and Publications, 1985.

7 Brzoska and Ohlson, op.cit., Figure 1.8, p. 9. It should be noted that

the figures here are cited for the 'Middle East', taken to include Iran and Israel but exclude North Africa and the Horn of Africa. Nonetheless, the ratios hold roughly true throughout.

8 *The Military Balance*, various editions.

9 American arms to Libya were mainly supplied either before Qaddafi's rule or in its early years.

10 In trend-indicator prices at 1985 constant prices.

11 On the British deals, *at-Tagrir*, Vol. 4, No. 14, 1–15 September 1988.

12 Brzoska and Ohlson, op. cit., 1988, Table 3.7, p. 113.

Chapter 3 Assessing Arab Defence Spending and Arms Imports

1 Conscription is authorised, but terms of service so far are voluntary.

2 Calculated from memo issued by National Association of Arab Americans (NAAA), 1986.

3 Local production would conceivably also save some of the money lost in paying commissions to middlemen for securing foreign arms contracts.

Chapter 4 Egypt: The Veteran Arab Arms Producer

1 *Al-Yom as-Sabi'* (Paris), 14 March 1988.

2 *Journal of Defence & Diplomacy*, (*Washington*) August 1986, p. 31.

3 Details on this entire period of development in Mohammad El-Sayed Selim, 'Egypt', in James Katz (ed.), *Arms Production in Developing Countries*, Lexington MA and Toronto: Lexington Books, 1984, pp. 123–6.

4 'Egyptian Defense Industry Directory', *Defense & Foreign Affairs*, June 1985, p. 20.

5 Op. cit., pp. 127–9.

6 Selim, op. cit., p. 132.

7 *Al-Ahram*, 2 July 1968.

8 *Al-Ahram*, 3 February 1987.

9 Mahmoud Riyad, *Mudhakkarat (Memoirs)*, Beirut: Arab Institute for Research and Publication, 1981, pp. 315–6.

10 Marouf Bakhit, *Egyptian Air Power in Transition, 1967–1973*, PhD thesis submitted to King's College London, in 1990. Unpublished.

11 Originally the AOI was known as the Arab Military Industry Organisation (AMIO).

12 R. Vayrynen, 'La lente construction de l'industrie militaire egyuptienne', *Le monde diplomatique*, October 1980.

13 General as-Sayyed mentions the aspect of resupply in particular, *Al-Ahram*, 3 February 1987.

14 As-Sayyed, ibid. On the Israeli side, see T.N. Dupuy, *Elusive Victory: The Arab-Israeli Wars, 1947–1974*, London: Macdonald & Jane's, 1978, Table F, p. 609.

15 *Al-Ahram*, 6 March 1974.

16 Vayrynen, op. cit.

17 *Al-Ahram*, 27 April 1974.
18 Thalif Deen 'Egypt set to head Arab arms bid', *Jane's Defence Weekly*, 9 April 1988, p. 685.
19 Nabil Ahmad, 'As-Sina'ah al-Harbiyyah al-Arabiyyah . . . Nadhrah Mustaqbaliyyah' ('The Arab Defence Industry: A Look to the Future'), Part I, *Al-Baheth al-Arabi* (London), No. 12, July-September 1987, p. 87; and R Vayrynen, 'La lente construction de l'industrie militaire egyptienne', *Le Monde Diplomatique*, October 1980.
20 According to Qatar's Sheikh Khalifah. *Al-Ahram*, 18 February 1978.
21 *Al-Ahram*, 8 September 1975.
22 *Al-Ahram*, 26 December 1977.
23 Kuna news agency; cited in *al-Watan*, 21 June 1978.
24 *Al-Ahram*, 2 December 1978.
25 R. Ropelewski, 'Arabs Seek Arms Sufficiency', *Aviation Week & Space Technology*, 15 May 1978; cited in Vayrynen, op. cit.
26 Ibid. and *Flight International*, 10 April 1977; 24 June 1978. Also cited in Vayrynen, op. cit.
27 *Al-Ahram*, 7 January 1979.
28 Quoted in Mahmoud 'Azmi, 'Khutwa Ula 'ala Tariq al-Sina'ah al-Harbiyyah al-'Arabiyyah ('A First Step on the Road to an Arab Defence Industry'), *Shu'un Filistiniyyah*, No. 78, May 1978, p. 189.
29 Vayrynen, op. cit.
30 *Al-Anba'*, 16 May 1979.
31 According to the current AOI head General Ibrahim al-'Orabi, interviewed in *al-Watan al-'Arabi*, 7 October 1988; and *Middle East* (London), No. 163, May 1988. Part of this sum was probably the $41 m profit made in 1976. Liberstein, 'Egyptian Defense Industry: Ambitious Plans', *Journal of Defence & Diplomacy*, August 1986, p. 30.
32 *Financial Times*, 30 September 1988. According to an AOI official, Egypt was counter-suing Westland for £80 million, claiming that the latter company had failed to meet contracts signed with the Organisation before July 1979. Mark Lambert, 'Egypt rebuilds its aircraft industry', *Interavia*, No. 2, 1984, p. 158.
33 Selim, op. cit., p. 133.
34 *Al-Akhbar*, 8 December 1987.
35 According to Major General Mohammad Farouq Mustafa, head of the Egyptian Armament Authority. *Al-Ahram*, 20 August 1988; and Thalif Deen, op. cit., p. 685.
36 For further information on the private sector in Egypt, see chapter 8.
37 Much of the information here is from 'Egyptian Defence Directory', loc. cit., pp. 20–1.
38 *Financial Times*, January 1988; and Jim Paul, 'The Egyptian Arms Industry', *Middle East Research and Information Project* (Washington, DC), February 1983, p. 28.
39 According to General as-Sayyed, *Aviation Week & Space Technology*, 4 January 1982, p. 41.
40 *Al-Jumhuriyyah*, 7 November 1985. The SIPRI estimate is somewhat lower, at $240 m.

41 According to General As-Sayyed, op. cit. in ref. 38.
42 *Ar-Ra'i al-Am* (Kuwait), 18 March 1987.
43 'Egypt: defence show gives military industry a boost', *Middle East Economic Digest*, 21 November 1987.
44 *Journal of Defense & Diplomacy*, August 1986, p. 32.
45 The Parliament decision taken on 4 July 1979; *Al-Ahram*, 5 July 1979.
46 *Al-Ahram*, 14 December 1979.
47 *Al-Jumhuriyyah*, 16 December 1983.
48 An American source sets the number of employees at 15,000. D. Louscher and M. Salome, *Technology Transfer and US Security Assistance: The Impact of Licensed Production*, Boulder and London: Westview Press, 1987, p. 124.
49 Selim, op. cit., p. 139.
50 *Al-Qabas* (Kuwait), 22 March 1988.
51 *Al-Ahram*, 10 March 1988.
52 *Middle East Economic Digest*, 2 February 1990.
53 *Aviation Week & Space Technology*, 4 January 1982, p. 45.
54 *Aviation Week & Space Technology*, 18 January 1982.
55 *Al-Ahram*, 29 January 1979.
56 M. Lambert, 'Egypt rebuilds its aircraft industry', *Interavia*, February 1984, p. 159
57 *Al-Ahram*, 3 March 1978.
58 Lambert, op. cit., p. 158.
59 *Middle East Economic Digest*, 26 January 1991.
60 *Middle East Economic Digest*, 11 May 1990 and 12 April 1991.
61 *Al-Quds*, 30 May 1989; *Financial Times*, 31 May 1989.
62 *Air International*, June 1982, pp. 282–83.
63 Ibid., April 1982, p. 201.
64 *Jane's Defence Weekly*, 9 April 1988, p. 657.
65 *Middle East Economic Digest*, 21 May 1982.
66 Robinson, 'Nation seeks . . .', op. cit., p. 44.
67 *Al-Musawwar*, No 3339, 7 October 1988.
68 *Al-Jumhuriyyah*, 7 November 1985; *Middle East Economic Digest*, 28 June 1986; *Al-Ahram*, 14 March 1988. The initial capital has also been stated at $11m, with an annual turnover of $25m by 1986.
69 In interview in *al-Musawwar*, No. 3339, 7 October 1988.
70 'Made in Egypt', *Defense & Foreign Affairs*, January 1988.
71 *Defense & Foreign Affairs*, October/November 1988.
72 *Jane's Defence Weekly*, 13 July 1985, p. 64.
73 *Al-Ahram*, 20 August 1988.
74 According to AOI head General Orabi; *Al-Ahram*, 21 December 1988.
75 Ahmad Heibeh, Strategic Group, AOI; *Al-Mustaqbal al-Arabi*, No. 124, June 1989, p. 110.
76 According to Qasim Ja'far, *Al-Ta'irat al-Qitaliyyah fi al-Mantaqah al-Arabiyyah: Ta'irat at-Tadrib wal-Musanadah (Combat Aircraft in the Arab Region: Training and Support Aircraft)*, Beirut: Arab Institute for Research and Publishing, 1985. However, Western sources suggest that only 36 *Gumhuriyyah* aircraft remain in service in Egypt.
77 Robinson, *AW&ST*, 4 January 1982, p. 44.

78 'Egypt: An Air Power in Transition', Part 2, *Air International*, May 1982, p. 278.
79 Robinson, *Aviation Week & Space Technology* 18 January 1982, p. 61.
80 *Middle East Research and Information Project*, 3/1983, p. 27.
81 *Interavia*, February 1984, p. 157.
82 *Al-Yom al-Sabi'*, Paris, 14 March 1988.
83 *Jane's Defence Weekly*, 22 April 1989.
84 *Jane's All the World's Aircraft 1966–1967*; though Robinson, op. cit., p. 42, asserts that designed speed was Mach 1.5.
85 Some Egyptian officials blame the closure on 'Soviet and American pressure' aimed at preventing Egypt from developing an independent ability. See, for example, General Nabeel Ibrahim, who charges that a German scientist called Heinkel sold the designs to France, which then developed the *Mystere* on their basis – a ludicrous claim, if only because the *Mystere* was already in service in the 1950s! *Al-Ittihad*, 2 October 1988.
86 *Aviation Week & Space Technology*, 18 January 1982.
87 'Egypt: An Air Power in Transition', Part 2, *Air International*, May 1982, pp. 241–2.
88 *Defense & Foreign Affairs*, November 1984, p. 1.
89 *Le Monde*, 16 August 1984.
90 *Al-Ahram*, 6 November 1985.
91 *Al-Qods al-Arabi*, 30 May 1989; *Financial Times*, 31 May 1989.
92 *Al-Anba*, 24 June 1979.
93 *Le Monde*, 7 June 1979.
94 *Defense & Foreign Affairs*, November 1984, p. 1.
95 'Egypt's Resurgent Aircraft Industry', *Jane's Defence Review*, Vol.3, No.2, March 1982, p. 267.
96 *Interavia*, February 1984, p. 158; and Clarence Robinson, 'Factories Tool for *Alpha Jet* Program', *Aviation Week & Space Technology*, 18 January 1982, p. 61.
97 Robinson, ibid, p. 61.
98 *Interavia*, February 1984, p. 157.
99 *Defense & Foreign Affairs*, November 1984, p. 1.
100 *Middle East Economic Digest*, 21 November 1987, p. 11.
101 *Middle East Economic Digest*, 18 November 1988.
102 *Al-Qods al-Arabi*, 28 April 1989.
103 *Shu'un Filistiniyyah*, No. 78, May 1978. Other figures cited are 280, in Vayrynen, *Le Monde*, and 230, in *Flight International*, 16 October 1978.
104 *Middle East Economic Digest*, 16 December 1988; and *Financial Times*, 30 September 1988.
105 Lambert, 'Egypt Rebuilds Its Defence Industry', *Interavia*, February 1984, p. 159.
106 Lambert, op. cit., p. 160.
107 *Jane's All the World's Aircraft 1987–1988*.
108 *MEED*, 21 November 1987, p. 11.
109 *Jane's Defence Weekly*, 21 November 1987.

110 'Cairo: Focus on Egyptian Programmes', *Military Technology*, February 1988, p. 82.
111 *Middle East Economic Digest*, 18 November 1988.
112 One is called *Najla. Jane's Defence Weekly*, 17 October 1987.
113 *Jane's Defence Weekly*, 21 November 1987.
114 According to AOI Vice-Chairman Eng. Nur Yusif. *Al-Ahram*, 5 October 1988.
115 *Jane's Defence Weekly*, 19 March 1988, p. 531.
116 For description of these tactics, see E. Luttwak and D. Horowitz, *The Israeli Army*, London: Allen & Unwin, 1975; and Mahmoud Azmi, *Al-Quwwat al-Muddarra'ah al-Isra'iliyyah Abra Arba' Huroub (The Israeli Armoured Forces through Four Wars)*, Beirut: PLO Research Center, 1975.
117 Estimates of the number of T-54/55s vary, with the same source suggesting 1,040 in 1988 (*Military Balance*, 1981–1982 and 1988–1989). The discrepancy appears to be due to better reporting.
118 Schaefer, '*Ramses* II: Egyptian modernization for the T-54', *International Defense Review*, February, 1988, p. 185.
119 *Military Technology*, October 1983.
120 *Jane's Armour and Artillery 1987–1988*, London: Jane's.
121 *Jane's Armour and Artillery 1983–1984*, London: Jane's, p. 819.
122 *Al-Ahram*, 29 April 1987.
123 *Ar-Ra'i al-'Am* (Kuwait), 29 April 1987.
124 Details in Schaefer, op. cit.
125 Foss, 'Egyptian tank upgrade "winners" ', *Jane's Defence Weekly*, Vol. 9, No. 20; 21 May 1988.
126 Schaefer, op. cit., p. 187.
127 *Middle East Economic Digest*, 9 November 1984.
128 *Defense and Foreign Affairs Weekly*, 12 November 1984.
129 *Middle East Economic Digest*, 9 November 1984.
130 *International Defense Review*, February, 1985, p. 213.
131 *Middle East Economic Digest*, 28 June 1986.
132 *Jane's Defence Weekly*, 16 January 1988.
133 *Al-Hawadeth*, 5 February 1988.
134 *International Defense Review*, February 1985.
135 *Al-Qabas*, 4 June 1986.
136 *Al-Ahram*, 29 April 1987.
137 *International Herald Tribune*, 30 June 1987.
138 *Military Technology*, February 1988, p. 80.
139 Ibid.
140 AOI officials dismissed claims that an Egyptian-produced M-1 would cost more than an imported one. Personal notes of seminar at the Center for Arab Unity Studies in Cairo, November 1988.
141 *International Herald Tribune*, 30 June 1987; and *Military Technology*, February 1988, p. 80. Moreover, the M-1 was offered to Saudi Arabia in 1989 at roughly $2.4m per tank, though follow-on support would double that figure.
142 *Jane's Defence Weekly*, 12 November 1988.
143 *Middle East Economic Digest*, 16 December 1988; and *Jane's Defence*

Weekly, 12 March 1988.
144 Egyptian claim, *Jane's Defence Weekly*, 12 November 1988; actual American terms in *Middle East Economic Digest*, 16 December 1988.
145 *Jane's Defence Weekly*, 30 July 1988.
146 *International Herald Tribune*, 30 June 1987.
147 *Jane's Defence Weekly*, 30 July 1987; and *Middle East Economic Digest*, 16 December 1988.
148 *Jane's Defence Weekly*, 30 September 1989.
149 *Al-Hayat*, 9 June 1989.
150 *Al-Hayat*, 13–14 and 17 May 1989.
151 He also accepted a higher level of Egyptian input into producing components for the fourth batch of F-16 combat aircraft that the country was to receive.
152 *Middle East Economic Digest*, 30 March 1990 and 25 January 1991.
153 *Middle East News Agency*, 3 October 1980.
154 *NATO's Sixteen Nations*, December 1984 – January 1985, p. 103–4.
155 *International Defence Review*, September 1987, p. 1240.
156 *International Defence Review*, February 1985, p. 217.
157 *African Defense*, March 1984.
158 *Jane's Defence Weekly*, 1 April 1989; and *Middle East Economic Digest*, 21 November 1987.
159 *International Defence Review*, February 1985, p. 215.
160 According to Abu Ghazalah. *BBC Summary of World Broadcasts*, 21 October 1980.
161 *Al-Qabas*, 13 August 1985.
162 *International Defence Review*, February 1985, pp. 214–5; and Lonscher & Salome, op. cit., p. 129.
163 *International Defence Review*, February 1985, p. 217.
164 *Aviation Week & Space Technology*, 4 January 1982, p. 45.
165 *Jane's Defence Weekly*, 12 December 1987.
166 *Aviation Week & Space Technology*, 4 January 1982.
167 Selim, op. cit., pp. 130–1.
168 'Military Industries in the Arab Region', *al-Mustaqbal al-'Arabi*, No. 124, 6/1989, June, p. 102.
169 *Military Balance 1988–1989*.
170 M. Navias, 'Ballistic missile proliferation in the Middle East', *Survival*, May/June 1989, p. 228.
171 *Al-Jumhuriyyah*, 8 November 1987.
172 Navias, op. cit., p. 228.
173 'Argentina, Egypt in Long-Range Missile Project', *Financial Times*, 21 December 1987.
174 *International Herald Tribune*, 21 and 25 September 1989.
175 Navias, op. cit., p. 228; and *International Herald Tribune*, 21 September 1989.
176 According to the head of the firm involved; *International Herald Tribune*, 27 June 1988.
177 D. Nelson, 'Ceausescu and the Romanian Army', *International Defense Review*, June 1989, p. 739.
178 *Jane's Defence Weekly*, 16 January 1988.

179 Ibid. and *al-Jumhouriyyah*, 8 November 1987.
180 *Aviation Week & Space Technology*, 4 January 1982.
181 *Jane's Defence Weekly*, 12 October 1985.
182 Hamdi Lutfi, 'Experts Develop New Mine Detector', *Middle East Times*, 10–16 April 1988.
183 *International Defense Review*, February 1985, p. 217; *Jane's Defence Weekly*, 20 May 1989.
184 *SIPRI Yearbook 1987*, p. 281.
185 *Al-Ahram*, 16 November 1985.
186 *International Defense Review*, February 1985, p. 219.
187 *Al-Hayat*, 18 May 1989.
188 *Al-Akhbar*, 12 January 1988.
189 *Aviation Week & Space Technology*, 4 January 1982.
190 *Yom al-Sabi'*, 14 March 1988.
191 According to ABDCo. head, Yusif Mustafa; *al-Ahram*, 16 November 1985.
192 *International Defense Review*, February 1985, p. 219: Part of the modification, such as installing the IFF unit, was done by Teledyne in the USA. 'Egypt: an Air Power in Transition', *Air International*, April 1982, p. 200.
193 *Middle East*, May 1988, p. 24.
194 *Oman*, 11 August 1985.
195 *Jane's Defence Weekly*, 10 September 1988.
196 *International Defense Review*, February 1985, p. 218.
197 *Financial Times*, 25 June 1988.
198 *Jane's Defence Weekly*, 17 November 1984.
199 According to AD Commander, *Middle East News Agency*, 25 June 1986.
200 *Oman*, 11 August 1985.
201 *NATO's Sixteen Nations*, December 1984–January 1985, p. 104.
202 *Jane's Defence Weekly*, 30 January 1988.
203 *Jane's Defence Weekly*, 12 November 1987.
204 *Jane's Defence Weekly*, 22 October 1988.
205 *Al-Hayat*, 19–20 August 1989.
206 According to Air Defence Commander General Adil Khalil, *Middle East News Agency*, 25 June 1986.
207 According to the Chairman of Benha's Board of Management, General Hasan Hasan. *Al-Ahram*, 18 October 1988.
208 *Middle East Economic Digest*, 21 November 1987.
209 The latter three companies belong to the Suez Canal Group of Companies.
210 'Egyptian Coast Guards', *Naval Forces*, Vol. IV, No. 3/1983; *Jane's Fighting Ships 1987–1988*, London.
211 *Jane's Fighting Ships 1987–1988*.
212 *International Defense Review*, February 1985, p. 222.
213 *Jane's Fighting Ships 1987–1988*.
214 *Jane's Defence Weekly*, 9 April 1988, p. 657.
215 *Jane's Defence Weekly*, 9 April 1988, p. 685.
216 'Iraqi ships to be overhauled in Egypt', *Egyptian Gazette*, 25 November 1988.

217 *Jane's Defence Weekly*, 17 October 1987, 9 April 1988.
218 Ahmad Heibah, Strategic Committee AOI, in *al-Mustaqbal al-'Arabi*, No. 124, p. 119.
219 Ibid.
220 *African Defence*, August 1986, p. 10.
221 It must be noted, however, that regardless of Saudi preferences, the USA has not extended permission to Egypt to make third-party sales of the M-1s it is scheduled to produce locally.
222 *Al-Hayat*, 16 June 1989.
223 *Al-Safir*, 10 November 1988.
224 Ibid. Yezid Sayigh, 'Arab regional security: between mechanics and politics', *RUSI Journal*, summer 1991.
225 These two projects were an Arab-Egyptian venture with a partial British loan, and a South Korean-Egyptian proposal, *al-Hayat*, 13 and 21 June 1989.
226 Robinson, 'Egypt: An Air Power in Transition', Part 3, p. 283. op. cit.
227 *African Defence*, April 1987, p. 10.
228 J. Paul, *Middle East Research and Information Project*, p. 26.
229 This has been the fate of the Nasr company, which produces Polish cars under licence; *al-Hayat*, 3 October 1989.
230 An aspect noted in M. Dunn, 'Egypt: from Domestic Needs to Export Market', in James Katz (ed.), *The Implications of Third World Military Industrialisation*, Lexington MA and Toronto: Lexington Books, 1986, p. 129.
231 Paul, op. cit., p. 2.

Chapter 5 Iraq: Emergent Arms Producer and Regional Power

1 G. Willis, 'Open Sesame: Baghdad show reveals Iraqi military-industrial capabilities', *IDR*, Vol. 22, No. 6, 1989, p. 836. International Defense Review.
2 Ali Qasim al-Oqabi, *Ath-Thawrah*, 11 October 1987.
3 According to the Deputy Minister for Industry and Military Industry, 'Behind the scenes in Baghdad', *International Defense Review*, June 1989, p. 841.
4 *International Herald Tribune*, 12 June 1989.
5 Though one Western economist offers a highly improbable estimate of \$100bn. *Al-Hayat*, 23 August 1989.
6 Direct physical damages have been unofficially estimated as high as \$35bn, besides lost oil revenues of \$15bn (*al-Hayat*, 23 August 1989). However, total losses have in fact been much higher, with a monthly war bill of \$600–1,000m, exhaustion of \$35bn of foreign exchange reserves held at the beginning of the conflict, and the loss of \$12–15bn in oil revenues *annually*. Goose, 'Armed conflicts in 1986, and the Iraq-Iran war', in *SIPRI Yearbook 1987*, Oxford: Oxford University Press, 1987, p. 299.
7 *Al-Ittihad*, cited in *Jane's Defence Weekly*, 22 April 1989.
8 *Al-Hayat*, 16 June 1989.
9 *Jane's Defence Weekly*, 13 May 1989.

10 *Al-Watan al-Arabi*, 28 October 1988.
11 *International Herald Tribune*, 12 June 1989.
12 *International Defense Review*, February 1985, p. 221.
13 *Military Technology*, July 1986.
14 *International Defense Review*, February 1985, p. 222.
15 In fact, these sources claim that between 4,668 and 4,860 barrels were produced in an unbelievable 25 days; *at-Tadamon*, 30 May 1987; *al-Dustour*, 25 May 1987.
16 *Jane's Defence Weekly*, 10 June 1989.
17 *Jane's Defence Weekly*, 20 May 1989.
18 *Jane's Defence Weekly*, 20 May 1989.
19 *Jane's Defence Weekly*, 13 May 1989.
20 *Jane's Defence Weekly*, 13 May 1989.
21 Willis, *International Defense Review*, June/1989, p. 836.
22 *Ad-Dustour*, 25 May 1987.
23 *Jane's Defence Weekly*, 20 May 1989.
24 Willis, op. cit, p. 837–8.
25 *Al-Hayat*, 21 April 1991; *Jane's Defence Weekly*, 28 September 1991.
26 *Le Point*, 27 March 1988; and Thalif Deen, 'Rebuilding Iraq's armed forces', *Jane's Defence Weekly*, p. 698.
27 *Sunday Times*, 27 January 1991.
28 *Jane's Defence Weekly*, 27 April 1991.
29 *Al-Watan al-'Arabi*, 28 October 1988 mentions the lower figures.
30 Willis, op. cit., p. 837.
31 *Jane's Defence Weekly*, 20 May 1989, p. 926.
32 *Jane's Defence Weekly*, 30 March 1991, p. 509.
33 'Rocket projects continue', *Jane's Defence Weekly*, 20 May 1989.
34 According to Sa'di interview. *International Defense Review*, June/1989, p. 841.
35 *Le Point*, 27 March 1988.
36 *Correspondent*, 8 October 1988.
37 *International Herald Tribune*, 5 July 1991.
38 *International Herald Tribune*, 24 November, 8 December 1989.
39 *Al-Hayat*, 26 November 1990.
40 *International Herald Tribune*, 21 September 1990.
41 *Arming Saddam: The Supply of British Military Equipment to Iraq 1979–1990*, London: Campaign Against Arms Trade, 1991, p. 13.
42 *Sunday Times*, 27 January 1991.
43 *Middle East*, December 1989, p. 29.
44 *Military Technology*, June 1990. Details also in *Middle East Economic Digest*, 27 April 1990.
45 *Jane's Defence Weekly*, 24 November 1990; Bull was murdered in Brussels in March 1990.
46 Details of these designs are in *Arming Saddam*, pp. 13–5.
47 Photographs and details in *Jane's Defence Weekly*, 14 September 1991, pp. 458–59.
48 *Reuters*, 8 October 1991.
49 'Iraqi T-55 MBT's greater firepower', *Jane's Defence Weekly*, 20 May 1989.

50 *Al-Hayat*, 13–14 May 1989.
51 *Military Technology*, July 1986.
52 *Jane's Defence Weekly*, 20 May 1989.
53 *Jane's Defence Weekly*, 20 May 1989; *International Defense Review*, June 1989, p. 837 and September 1991, pp. 996–97.
54 *International Defense Review*, June 1989, p. 837.
55 *International Defense Review*, June 1989, p. 837.
56 'Appliqué armour, 30-mm gun for BMP-1', *Jane's Defence Weekly*, 13 May 1989.
57 *International Herald Tribune*, 12 June 1989.
58 'Iraq equips *Mirages* with *Kedge* missiles', *Jane's Defence Weekly*, 8 July 1989.
59 *Jane's Defence Weekly* 4 November 1989.
60 'Rocket projects continue', *Jane's Defence Weekly*, 20 May 1989.
61 Ibid; *International Defense Review*, June 1989, p. 838.
62 Ibid., June 1989, p. 837.
63 Ibid. p. 838.
64 *Financial Times*, 30 August 1990.
65 *International Defense Review*, June 1989, p. 841.
66 Timmerman, *International Herald Tribune*, 12 June 1989.
67 The British *Hawk* would have been a strong competitor, but Iraqi officials demanded political guarantees from the British government concerning the continuity of the programme and of other related arms transfers and licences; Britain subsequently prohibited the proposed sale.
68 ' "Candid" AEW one of Baghdad show surprises', *Jane's Defence Weekly*, 13 May 1989.
69 *International Herald Tribune*, 12 June 1989.
70 *Jane's Defence Weekly*, 13 May 1989.
71 *Jane's Defence Weekly*, 2 September 1989.
72 *Jane's Defence Weekly*, 7 October 1989. A definite innovation is the addition of two strakes on the lower rear fuselage to provide lateral counterbalance to the effect of the rotodome.
73 *International Defense Review*, June 1989, p. 838.
74 *Jane's Defence Weekly* 13 July 1991, p. 58, and October 1991, pp. 402–3.
75 'Iraq now Middle East's biggest chemical weapon producer'. *Jane's Defence Weekly*, 27 February 1988.
76 *Arming Saddam*, p. 15.
77 *Al-Hayat*, 21 April 1991; *Jane's Defence Weekly*, 27 April 1991.
78 *International Herald Tribune*, 9 July 1991.
79 *Al-Hayat*, 3 May 1991.
80 On the smuggling of nuclear components from the West, see *International Herald Tribune*, 19 November 1990. A list of the main nuclear centres and their stocks and of the centrifuge effort is in *Le Monde*, 30 August 1990.
81 *Jane's Defence Weekly*, 16 March 1991, p. 407, and 13 April 1991, p. 611; and *International Defence Review*, June 1991, pp. 567–70.
82 *Financial Times*, 2 December 1990.

83 Ibid., 21 September 1989.
84 British newspaper reports cited in *Arming Saddam*, p. 11.
85 *Middle East*, December 1989, p. 29.
86 *Financial Times*, 10 November 1990.
87 Ibid., 10 November 1990.
88 *Middle East Research and Information Project*, January-February 1991, p. 30.
89 *Sunday Times*, 27 January 1991.
90 *Financial Times*, 31 March 1990; *Defence*, June 1990, cited in *Arming Saddam*, p. 13.
91 *Middle East Research and Information Project*, January-February 1991, p. 30.
92 *Middle East*, December 1989.
93 *Middle East Research and Information Project*, January-February 1991, p. 30.
94 Ibid., p. 30. Additional information on the German role is in the *International Herald Tribune*, 7 February 1991.
95 *Sunday Times*, 16 December 1990.
96 *Sunday Times*, 15 July 1990; *Arming Saddam*, p. 17.
97 22 April 1989. *Jane's Defence Weekly*.
98 According to General Ammar, Head of the Scientific and Development Committee; interview in *Military Technology*, July 1986.
99 22 April 1989. *Jane's Defence Weekly*.
100 19 May 1989, p. 2. *Middle East Economic Digest*.
101 *Jane's Defence Weekly*, 13 April 1991.
102 Ibid., 1991.
103 *International Herald Tribune*, 27 June 1991.
104 *Al-Hayat*, 3 May 1991.

Chapter 6 Saudi Arabia: One Foot Forward, One Foot Back

1 *International Herald Tribune*, 17 December 1983.
2 *Financial Times*, 20 August 1984.
3 *Middle East Economic Digest*, 31 October 1987.
4 *Defence Industries of the Middle East*, Vol. 1, No. 1, Summer 1988, London: Gulf Centre for Strategic Studies, p. 57.
5 *Financial Times*, 20 August 1984.
6 *Middle East Economic Digest*, 31 October 1987.
7 *International Herald Tribune*, 1 October 1984.
8 Nabeel Ibrahim, 'Arab Defense Industries: A Futuristic Look' Part 3, *al-Baheth al-Arabi*, 1 January to March/1988, p. 94. The first two items may have been slated for Saudi funding during development only, while the *Tucano* was to be assembled in Saudi Arabia.
9 Details in A. Cordesman, *The Gulf and the West: Strategic Relations and Military Realities*, Boulder and London: Westview and Mansell, 1988, p. 262.
10 *Defence Industries of the Middle East*, pp. 57, 58.
11 *Financial Times*, 22 October 1987.

12 *Middle East Economic Digest*, 22 July 1988.
13 'UK "Al Yamamah" Economic Offset Programme', Ministry of Defence News Release, 23 November 1988.
14 *Financial Times*, 22 October 1987. Later reports suggested that the share was 25 per cent of British technical component in the total £15bn deals.
15 *Offset Memo*.
16 *Middle East Economic Digest*, 16 December 1988.
17 *Financial Times*, 22 October 1987.
18 *Hawadeth*, 11 February 1986.
19 Ibid.
20 *Defence Industries of the Middle East*, p. 59. Some 850 family houses and 400 bachelor apartments were planned; *Hawadeth*, 11 February 1986.
21 According to then Director-General of GEMI, General Yusif Ibrahim Salloum; *Sharq al-Awsat*, 30 January 1986.
22 *Defence Industries of the Middle East*, p. 60.
23 *Jane's Defence Weekly*, 10 June 1989.
24 *Middle East Economic Digest*, 31 October 1987.
25 *Defence Industries of the Middle East.*
26 *Hawadeth*, 11 February 1986.
27 *Financial Times*, 17 January 1989.
28 *International Defence Review*, editorial supplement to March 1989 issue, *Defense Electronics & Computing*, No. 1, p. 22.
29 *International Herald Tribune*, 17 December 1983.
30 *Defense & Foreign Affairs*, March 1984.
31 *International Herald Tribune*, 17 December 1983.
32 *Middle East Economic Digest*, 22 July 1988.
33 *Financial Times*, 24 November 1988.
34 *Middle East Economic Digest*, 31 October 1987.
35 *Middle East Economic Digest*, 16 December 1988.
36 *Middle East Economic Digest*, 22 July 1988.
37 This is implied in A. Cordesman, *The Gulf and the Search for Strategic Stability: Saudi Arabia, the Military Balance in the Gulf, and Trends in the Arab-Israeli Military Balance*, Boulder and London: Westview and Mansell, 1984, pp. 354–55.
38 *Defense & Foreign Affairs Weekly* 11–17 March 1985; Cordesman, 1988, op. cit., ref. 9, p. 262.
39 *Middle East Economic Digest*, 22 July 1988.

Chapter 7 Other Arab Countries

1 *Defense & Foreign Affairs*, March 1984.
2 A description of the Rabta complex is in *The Observer*, 15 January 1989.
3 *The Correspondent*, 8 and 15 October 1989.
4 *Middle East Economic Digest*, 5 March 1988.
5 According to a senior member of Fateh's 'Scientific Committee', interviewed in Tunis in February 1989. He also confirms that similar co-

operation has been suggested by both Libya and Algeria at various times.

6 'Iraq now Middle East's biggest chemical weapon producer', *Jane's Defence Weekly*, 27 February 1988.

7 *Jane's Armour and Artillery 1983–1984.*

8 The fact that arms and ammunition could be bought cheaply from the USSR was cited by the Syrians, to the PLO, as a reason for suspending proposed production ventures. See note 5.

9 *Defence Industries of the Middle East*, p. 11. The licence is alternatively reported to be from Britain, *Middle East Economic Digest*, 31 October 1987.

10 *Jane's Fighting Ships 1987–1988.*

11 *Defence Industries of the Middle East*, p. 10.

12 Ibid.

13 Z.Y. Hershlag, 'Industrialisation in Arab Countries: Patterns, Options and Strategies', in R Aliboni (ed.), *Arab Industrialisation and Economic Integration*, London: Croom Helm, 1979, p. 44.

14 A Cordesman, *Jordanian Arms and the Middle East Balance*, Washington DC: Middle East Institute, 1983, p. 63.

15 *Jane's Defence Weekly*, 2 September 1989, p. 410.

16 'Jordan's crisis spurs rebuilt arms industry', *Jane's Defence Weekly*, 16 September 1989.

17 *Jane's Defence Weekly*, 16 September 1989, p. 501.

18 *Jane's Defence Weekly*, 2 September 1989, p. 410.

19 *International Defense Review*, August 1989, p. 1060.

20 *International Defense Review*, August 1989, p. 1060.

21 *International Defense Review*, August 1989, p. 1060.

22 *Jane's Defence Weekly*, 16 September 1989, p. 501.

23 The information here is based on interviews with two senior members of the committee.

Chapter 8 Technical and Operational Assessment of Arab Military Industry

1 Ahmad Heibeh cited in a panel discussion, 'al-Sina'at al-'Askariyyah fil-Watan al-'Arabi (Arab Military Industries)', *al-Mustaqbal al-'Arabi* (Beirut), No. 124, June 1989, p. 120.

2 Ibid., p. 120.

3 Catrina, *Arms Transfers and Dependence*, London: Taylor & Francis for UNIDIR, 1988, p. 193.

4 *Jane's Defence Weekly*, 30 July 1987; *Middle East Economic Digest*, 16 December 1987.

5 This assumes that collaboration has been suspended since the Gulf crisis, though this may not, in fact, be the case.

6 Based on an estimated $300m of military production in 1987. Abu Ghazalah, *al-Ra'i al-'Am* (Kuwait), 18 March 1987; and *Yom al-Sabi'*, 14 March 1988.

7 Ferrous and non-ferrous castings, forgings and sections; irrigation

equipment; reinforced aluminum cables for high tension lines; industrial explosives for mining; chemicals for detergents, lacquers and paints; rubber parts for industrial purposes; laboratory and medical ether; petrol and diesel engine parts: bodies, pistons and piston rings, piston liners, camshafts, cranks, radiators, ball bearings; tightening equipment: bolts and nuts for industry and railways (*Military Technology*, No. 9/1986, p. 45, Annex No. 2).

8 Interviews with the Jordanian Crown Prince Hasan bin Talal and an Air Force general, March 1984. Also see *Jordan Yearbook 1972* (in Arabic).

9 *Yom al-Sabi'*, 14 March 1988.

10 A. Zahlan, *al-' Arab wal-'Ilm wat-Teknolojia (The Arabs. Science and Technology)*, Beirut: Center for Arab Unity Studies, 1988. p. 49.

11 H. Carnegy and T. Walker. 'De facto disarmament', *Financial Times*, 4 September 1989.

12 Here total output refers to military goods only, excluding civilian products made by the defence industry, since the latter are not included in export figures.

13 Official Israeli figures on exports and sales in *Jane's Defence Weekly*, 24 June 1989. Discussed in Yezid Sayigh, 'Israeli Military Industry: Towards a Revival?', *Tagrir Askari*, September 1989.

14 *SIPRI Yearbook 1986*, p. 338.

15 18 January 1988. *Financial Times*.

16 *Safir*, 10 November 1988.

Chapter 9 Assessing the Economic Impact

1 *Unified Arab Economic Report*, 1987. Abu Dhabi: Arab Monetary Fund, n.d., Table 3. p. 61. This figure includes employment in the electricity and the gas industry, and certain branches of construction and civil engineering.

2 Figures from the *Unified Arab Economic Report, 1989*, Abu Dhabi: Arab Monetary Fund, n.d., Appendix 20/3, pp. 280–1.

3 By way of contrast, Singapore, which hopes for a GDP growth rate of 4–6 per cent annum in the 1990s, can only compensate for the low growth rate of its workforce (1.2 per cent p.a.) by raising productivity and increasing automation.

4 *World Bank Tables 1991*, Oxford: Oxford University Press for the World Bank, 1991.

5 *World Bank Tables 1989 and 1991.*

6 *World Bank Tables 1991* and *Military Balance 1990–91*.

7 *Unified Arab Economic Report 1989*. Abu Dhabi: Arab Monetary Fund, n.d., Appendix 14/3, p. 274.

8 A. Hewedy, *Militarization and Security in the Middle East*, London and New York: Pinter and St Martin's Press, 1989, p. 73.

9 Figures from *al-Hayat*, 11 August 1989: *Jane's Defence Weekly*, 24 June 1989.

Chapter 10 Arab Scientific and Industrial Base

1 C. Catrina. *Arms Transfers and Dependence*, London: Taylor & Francis for UNIDIR, 1988, p. 193.
2 Abu Dhabi: Arab Monetary Fund, n.d., Table 2, p. 59.
3 *Unified Arab Economic Report 1987*. Abu Dhabi: Arab Monetary Fund, n.d., Appendices 1/2 and 2/4. These figures have been used rather than the World Bank tables for consistency. In any case they do not differ substantially.
4 *World Bank Tables 1989*, Oxford: Oxford University Press for the World Bank, 1989.
5 *Al-Ahram*, 5 November 1988.
6 *Al-Hayat*, 5 October 1991.
7 *Middle East Economic Digest*, 5 May and 21 July 1989 and 23 February 1990.
8 Local input is probably highest in Egypt, Local input has accounted for 30% of production on the *Antar* tractor. *Middle East Economic Digest*, 5 May 1989.
9 According to AOI Chairman General Orabi, who added that the initial investment would cost $60m. *Al-Ahram*, 14 March 1988.
10 *Middle East Economic Digest*, 19 May 1989.
11 *Al-Hayat*, 13 June 1989. At the end of 1989. Egypt was still searching for additional private capital to launch the project. Capital was set at $50m and initial investment at $68m.
12 *Catalogue of Joint Arab Ventures*, Arab League and OAPEC, 1984.
13 Ibid.
14 Z. Y. Hershlag. 'Industrialisation in Arab Countries: Patterns, Options and Strategies', in R. Aliboni (ed.), *Arab Industrialisation and Economic Integration*. London: Croom Helm, 1979, p. 72.
15 *Catalogue of Joint Arab Ventures*.
16 J. S. Birks and C. A. Sinclair. *Arab Manpower: The Crisis of Development*, London: Croom Helm, 1980, pp. 222–3.
17 Op. cit., p. 67.
18 Abu Dhabi: Arab Monetary Fund, n.d., p. 69.
19 A. Zahlan, *The Arabs, Science and Technology*, Beirut: Centre for Arab Unity Studies, 1988, p. 49.
20 *Science and Science Policy in the Arab World*, Beirut: Centre for Arab Unity Studies, 1979 (4th edn, 1984), Table 2–3, p. 35.
21 Ibid., p. 40.
22 Ibid., p. 40.
23 Ibid., Table 2–5, p. 39. and p. 40.
24 A critical and up-to-date argument on Arab scientific output is Antoine Zahlan: 'The Challenge and Response: the Contribution of Arab Sciences and Technology to Modernization in the Arab homeland', *al-Mustaqbal al-'Arabi*, April 1991.
25 Subhi al-Qasim *et al.*, *Istratijiyyat Tatwir al-'Ulum wat-Taqanah fil-Watan al-'Arabi (Strategy for Developing the Sciences and Technology in the Arab World)*. Beirut: Centre for Arab Unity Studies and ALECSO. 1989, pp. 45, 47.

26 Ibid., p. 48.
27 Ibid., Table 11, pp. 46–7, 49.
28 Cited in Zahlan, 1979; calculated from Table 5–1, p. 98.
29 Ibid., p. 99.
30 Oasim op. cit., p. 50.
31 Figures here are for 1981–84, according to country, Ibid., Table 15, p. 52.
32 Ibid., Table 15, p. 52.
33 Ibid., Table 13, p. 50.
34 Hershlag, op. cit., p. 61.
35 Qasim, op. cit., p. 11.
36 Ibid., Table 2, p. 12.
37 According to the head of the Network of Science Academies in the Third World, Mohammad Abdul-Salam, *al-Hayat*, 7 September 1989.
38 Ibid., p. 185.
39 Aliboni, op. cit., p. 54.
40 G. Kennedy, *The Military in the Third World*, London: Duckworth, 1974, p. 295.
41 According to A. Zahlan, private discussion.
42 *Unified Arab Economic Report, 1987*, Abu Dhabi; Arab Monetary Fund, n.d., Appendix 13/17, p. 341; *Jordan Times* (Amman), 3 March 1986.
43 Report on Technology and National Security and Military Industry cited in *al-Hayat*, 28 September 1989.
44 Qasim, op. cit., pp. 184, 199.
45 *Unified Arab Economic Report 1987*, Abu Dhabi: Arab Monetary Fund, n.d., Appendix 3/4, p. 291.
46 Report on Technology and National Security and Military Industry cited in *al-Hayat*, 28 September 1989.
47 Dr Nabil Sha'ath, Director of the Consultancy firm TEAM interviewed in *Filastin al-Thawrah*, 19 November 1989.
48 Based on his presentation in 'Developing Countries', in N. Ball and M. Leitenberg (eds.), *Structure of the Defence Industry*, London: Croom Helm, 1983, p. 312.
49 Ibid. p. 324.

Chapter 11 Issues and Determinants of Future Military Industrialisation

1 *Al-Arabi*, December 1985, p. 18.
2 *Egyptian Gazette*, 22 November 1988.
3 Qasim, op.cit., p. 180.
4 *Jane's Defence Weekly*, 13 May 1989, p. 845.
5 M. Kaldor and A. Eide *eds., The World Military Order: The Impact of Military Technology on the Third World*, London: MacMillan, 1979.
6 Hormuz Mama, 'India's light combat aircraft'. *International Defense Review*, September 1989, p.1198.
7 Subhi al-Qasim et al., *Strategy for Developing the Sciences and*

Technology in the Arab World, Beirut: Centre for Arab Unity Studies and ALESCO, 1989.

8 According to the head of the Network of Science Academies in the Third World. Dr M. Abdul Salam, *al-Hayat*. 7 September 1989.

9 Saadet Deger, *Military Expenditure in Third World Countries*, London: Routledge 1986, Table 7.6, p. 173.

10 Report on Technology and National Security and Military Industry, cited in *al-Hayat*. 28 September 1989.

11 J. Nolan, *Military Industry in Taiwan and South Korea*. London: Macmillan, 1986, p.49.

12 Qasim, *op cit*, p. 184.

13 *Unified Arab Economic Report 1987*, Abu Dhabi: Arab Monetary Fund, n.d., p. 341.

14 Ibid., p. 185.

15 Ibid., p. 67.

16 Saudi Arabia has achieved greater results in its petroleum and petrochemicals sector, it is true, but remains over-dependent on foreign input generally.

17 For comparison, although Brazil has a generally open economy, the government (the Army, in fact) holds sole control over the computer sector and the transfer of computer technology, while permitting full private participation in it.

18 Deger, *op. cit.*, p. 158.

19 *Military Technology*, September 1986, p. 43.

20 Catrina, *Arms Transfers and Dependence*, London: Taylor & Francis for UNIDIR, 1988, p. 189.

21 P.R. Chari, 'Indo-Soviet Military Co-operation: a Review', *Strategic Digest*, May 1979, p. 303; cited in H. Wulf, 'Developing Countries', in N. Ball and M. Leitenberg, *The Structure of the Defence Industry: an International Survey*, London: Croom Helm, 1983, p. 331.

22 R. Looney, *Third World Military Expenditure and Arms Production* London: Macmillan, 1988, p. 112.

23 *Financial Times*, 3 February 1989.

24 *Jane's Defence Weekly*, 24 June 1989.

25 H. Wulf, op. cit., p. 320.

26 Catrina, op. cit., p. 192.

27 A. Ross, 'World Order and Third World Arms Production', in Katz (ed.), *The Implications of Third World Military Industrialization*, Lexington MA: Lexington, 1989, p. 279.

28 *Al-Ahram*, 21 December 1988.

29 On the disadvantages of this prospect, see Nolan, op. cit. p. 8.

30 Op. cit., p. 112.

31 Op. cit., p. 103.

32 AOI Strategic Committee, Ahmad Heibeh. 'al-Sina'at al-'Askariyyah fil-Watan al-'Arabi' ('Military Industries in the Arab Region'), *al-Mustaqbal al-'Arabi* (Beirut), June 1989, p. 119.

33 Deger, op. cit., p. 156.

34 *The Baroque Arsenal*, London: Andre Deutsch, 1981, pp. 226–30.

Chapter 12 Towards an Alternative Strategy for Arab Military Industrialisation

1 The UN Security Council resolution passed in April 1991 banned the supply of anything relating to the design, development, manufacture, and use of weapons, among other restrictions on NBC and ballistic missile programmes (*Jane's Defence Weekly*, 13 April 1991). The US Treasury meanwhile compiled its own blacklist of companies and persons said to have assisted 'Iraq's subterranean network in the world of arms, trading and clandestine financial operations' (Ibid.).
2 Subhi al-Qasim *et al.*, *Strategy for Developing the Sciences and Technology in the Arab World*, Beirut: Centre for Arab Unity Studies and ALECSO, 1989, p. 185.
3 Qasim, ibid., p. 207.
4 Report on Technology and National Security and Military Industry, cited in *al-Hayat*, 28 September 1989.
5 This point was stressed by Zahlan in a private interview.
6 This is a new field of study, discussed with limited success in Cimbala (ed.), *Information and Foreign Policy*, Lexington, MA: Lexington, 1988.
7 Allan Din *ed.*, *Arms and Artificial Intelligence*, Oxford: Oxford University Press, 1987, p. 3.
8 For fuller, technical discussions, see, for example: William Kincade, 'New military capabilities: Propellants and implications', and Phil Williams, 'Emerging technology, exotic technology and arms control', in Carl Jacobsen *ed.*, *The Uncertain Course: New Weapons, Strategies and Mind-Sets*, Oxford: Oxford University Press, 1987. A discussion of current military technology is: Major General Ken Perkins *ed.*, *Weapons & Warfare: Conventional Weapons and their Roles in Battle*, London: Brassey's, 1987.
9 *Financial Times*, 27 June 1988; and *Third World Defence*, 7 December 1988, p. 66.
10 *African Defence*. 8/1986.
11 Qasim, *op cit*, p. 66.
12 National Association of Arab-Americans Memo, 1988.
13 *Jane's Defence Weekly*, 21 October 1989.
14 *Al-Ahram*, 3 February 1987.
15 *Al-Anba*, 5 December 1987.
16 Catrina, op. cit., pp. 228–9.
17 Catrina, op. cit., Table 12–2, p. 231.
18 Neuman, op. cit. 1981, pp. 137–39.
19 *Jane's Defence Weekly*, 31 May 1986, p. 973.
20 Catrina, op. cit., p. 232.

Chapter 13 Epilogue

1 This argument is developed in Yezid Sayigh. 'Arab regional security: between mechanics and politics', *RUSI Journal*, Summer 1991.

2 Secretary of State Baker made American commitment to Israeli superiority clear in his testimony before the House Foreign Affairs Committee on 4 September 1990.

3 These patterns are discussed in Yezid Savigh, 'The Gulf crisis: why the Arab regional order failed'. *International Affairs*, July 1991.

4 *International Herald Tribune*, 23 March 1991.

Select Bibliography

Books

Roberto Aliboni (ed.), *Arab Industrialisation and Economic Integration*, London: Croom Helm, 1979.

Kwang-Il Baek, Ronald McLaurin, and Chung-In Moon (eds.), *The Dilemmas of Third World Defense Industries*, Boulder Co and Inchon: Westview Press and CIS-Inha University, 1989.

Bishara Bahbah, *Israel and Latin America and the Military Connection*, London: Mamillan, 1986.

Nicole Ball and Milton Leitenberg, *The Structure of the Defence Industry: An International Survey*, London: Croom Helm, 1983.

Hazem Beblawi and Giacomo Luciani (eds.), *The Rentier State*, London: Croom Helm, 1987.

J.S. Birks and C.A. Sinclair, *Arab Manpower: The Crisis of Development*, London: Croom Helm, 1980.

Michael Brzoska and Thomas Ohlson, *Arms Transfers to the Third World 1971–85*, Oxford: Oxford University Press for SIPRI, 1988.

Michael Brzoska and Thomas Ohlson (eds.), *Arms Production in the Third World*, London: Taylor & Francis for SIPRI, 1986.

Christian Catrina, *Arms Transfers and Dependence*, London: Taylor & Francis for UNDIR, 1988.

Saadet Deger, *Military Expenditure in Third World Countries*, London: RKP, 1986.

Amin Hewedy, *Militarization and Security in the Middle East: Its Impact on Development and Democracy*, London: Pinter Publishers, 1989.

Amin Hewedy, *Sina'at al-Aslihah fi Isra'il (Arms Production in Israel)*, Cairo: Dar al-Mustaqbal al-'Arabi for the General Bureau of Economic Affairs (League of Arab States), 1986.

Russell Howe, *Weapons*, London: Abacus, 1981.

Jane Hunter, *Israeli Foreign Policy*, Nottingham: Spokesman for the Bertrand Russell Peace Foundation, 1987.

Paul Jabber, *Not By War Alone*, Berkeley: California University Press, 1981.

Mary Kaldor, *Baroque Arsenal*, London: Andre Deutsch, 1982.

Mary Kaldor and Asbjorn Eide, *The World Military Order: The Impact of Military Technology on the Third World*, London: Macmillan, 1979.

James Katz (ed.), *Arms Production in Developing Countries*, Lexington MA and Toronto: Lexington Books, 1984.

James Katz (ed.), *The Implications of Third World Military*

Industrialization: Sowing the Serpent's Teeth, Lexington MA and Toronto: Lexington Books, 1986.

Gavin Kennedy, *The Military in the Third World*, London: Duckworth, 1974.

Michael Klare and Cynthia Arnson (et al), *Supplying Repression*, Washington DC: Institute for Policy Studies, 1981.

Aharon Klieman, *Israel's Global Reach*, Washington DC and London etc: Pergamon/Brassey's, 1985.

Signe Landgren, *Embargo Disimplemented: South Africa's Military Industry*, Oxford: Oxford University Press for SIPRI, 1989.

Ariel Levite and Athanassios Platias, *Evaluating Small States' Dependence on Arms Imports: An Alternative Perspective*, Cornell University, Peace Studies Program, Occasional Papers, No. 16.

Robert Looney, *Third World Military Expenditure and Arms Production*, London: Macmillan, 1988.

David Louscher and Michael Salome, *Technology Transfer and US Security Assistance: The Impact of Licensed Production*, Boulder and London: Westview Press, 1987.

Tal'at Musallam, *al-Ta'awun al-'Askari al-'Arabi (Arab Military Cooperation)*, Beirut: Center for Arab Unity Studies, 1990.

Stephanie Neuman, *Military Assistance in Recent Wars: The Dominance of the Superpowers*, Washington DC: Praeger for CSIS, The Washington Papers, No. 122, 1986.

Stephanie Neuman, *Defense Planning in Less-Industrialized States*, Lexington MA and Toronto: Lexington Books (Heath), 1984.

Janne Nolan, *Trappings of Power: Ballistic Missiles in the Third World*, Washington DC: Brookings, 1991.

Janne Nolan, *Military Industry in Taiwan and South Korea*, London: Macmillan, 1986.

Yoram Peri and Amnon Neubach, *al-Mujamma' al-'Askari as-Sina'i fi Isra'il (The Military-Industrial Complex in Israel)*, Beirut: Institute for Palestine Studies, 1985.

Subhi al-Qasim (et al.), *Istratijiyyat Tatwir al-'Ulum wat-Taqanah fil-Watan al-'Arabi (Strategy for Developing the Sciences and Technology in the Arab World)*, Beirut: Center for Arab Unity Studies and ALECSO, 1989.

Stewart Reiser, *The Israeli Arms Industry: Foreign Policy, Arms Transfers, and Military Doctrines of a Small State*, New York and London: Holmes & Meier, 1989.

Mahmoud Riyad, *Mudhakkarat (Memoirs)*, Beirut: Arab Institute for Research and Publishing, 1981.

Marek Thee (ed.), *Arms and Disarmament: SIPRI Findings*, Oxford and New York: Oxford University Press for SIPRI, 1986.

David Whynes, *The Economics of Third World Military Expenditure*, London: Macmillan, 1979.

Antoine Zahlan, *al-'Arab wal-'Ilm wat-Teknolojia (The Arabs, Science and Technology)*, Beirut: Center for Arab Unity Studies, 1988.

Antoine Zahlan, *al-'Ilm was-Siyasah al-'Ilmiyyah fil-'Alam al-'Arabi (Science and Science Policy in the Arab World)*, Beirut: Center for Arab

Unity Studies, 1979 (4th edition 1984).
Antoine Zahlan, *al-Bu'd at-Teknoloji lil-Wihdah al-'Arabiyyah (The Technological Dimension of Arab Unity)*, Beirut: Center for Arab Unity Studies, 1981 (1st edition 1983).

Journal articles and book chapters

(Individual newspaper and magazine articles are not cited)

(No author given) 'al-Sina'at al-'Askariyyah fil-Watan al-'Arabi (Military Industries in the Arab Region)', *Al-Mustaqbal al-'Arabi* (Beirut), No. 124, June 1989.
Michael Brzoska and Thomas Ohlson, 'Arms production in the Third World: an overview', in Michael Brzoska and Thomas Ohlson (eds.), *Arms Production in the Third World*, London: Taylor & Francis for SIPRI, 1986.
Michel Chatelus, 'Problems of Arab Industry', in Hazem Beblawi and Giacomo Luciani (eds), *The Rentier State*, London: Croom Helm, 1987.
Z.Y. Hershlag, 'Industrialisation in Arab Countries: Patterns, Options and Strategies', in Roberto Aliboni (ed.), *Arab Industrialisation and Economic Integration*, London: Croom Helm, 1979.
Amin Hewedy, 'Hal al-Tasni' al-Harbi Wasilah li Tahrir al-Qarar al-Siyasi? (Is Military Industrialization a Means to Liberate Political Decision-Making?)', *al-Arabi*, No. 325, December 1985.
Nabeel Ibrahim, 'As-Sina'ah al-Harbiyyah al-Arabiyyah: Nadhrah Mustaqbaliyyah (Arab Defence Industries: A Futuristic Look)', *al-Baheth al-Arabi*, Parts 1–3, Nos. 12–14, 1987 and 1988.
Aaron Karp, 'Ballistic Missiles in the Third World', *International Security*, Vol. 9, No. 3, Winter 1984/85.
Michael Klare, 'The Unnoticed Arms Trade: Exports of Conventional Arms-Making Technology', *International Security*, Vol. 8, No. 2, Fall 1983.
Martin Navias, 'Ballistic missile proliferation in the Middle East', *Survival*, May/June 1989, Vol. XXXI, No. 3.
Stephanie Neuman, 'Co-production, barter, and countertrade: offsets in the international arms market', *Orbis*, Vol. 29, No. 1, Spring 1985.
Andrew Ross, 'World Order and Third World Arms Production', in Katz (ed.), *The Implications of Third World Military Industrialization*, Lexington MA: Lexington, 1989.
Yezid Sayigh, 'Arab regional security: between mechanics and politics', *RUSI Journal*, Summer 1991.
Yezid Sayigh, 'Israeli Military Industry: Towards a Revival?', *Taqrir Askari*, September 1989. (In Arabic.)
Will Schaefer, 'Ramses II: Egyptian modernization for the T-54', *International Defense Review*, Vol. 21, No. 2, 1988.
Raimo Vayrynen, 'The Arab Organization of Industrialization: A Case Study in the Multinational Production of Arms', *Current Research on Peace and Violence*, 1979.
Herbert Wulf, 'Developing Countries', in Nicole Ball and Milton

Leitenberg, *The Structure of the Defence Industry: An International Survey*, London: Croom Helm, 1983.

Antoine Zahlan, 'al-Tahaddi: Musahamat al-'Ulum wat-Taqanah al-'Arabiyyah fi Tahdith al-Watan al-'Arabi' ('The Challenge: The Contribution of Arab Sciences and Technology in Modernising the Arab Homeland'), *al-Mustaqbal al-'Arabi*, No. 146, April 1991.

Yearbooks and reference works

Catalogue of Joint Arab Ventures, Arab League and OAPEC, 1984.

Jane's All the World's Aircraft

Jane's Armour and Artillery

Jane's Fighting Ships

SIPRI Yearbook: World Armaments and Disarmament, Oxford: Oxford University Press for SIPRI, successive years.

The Military Balance, London: IISS, successive years.

The Middle East Military Balance, Tel Aviv: JCSS, successive years (since 1983).

Kasim Ja'far, *Mizan al-Qiwa al-'Askariyy fi Mintaqat al-Sharq al-Awsat, 1984–1985 (The Balance of Military Forces in the Middle East Region, 1984–1985)*, Beirut: Arab Institute for Research and Publications, 1985.

World Bank Tables

Unified Arab Economic Report

Newspapers and magazines

Al-Hayat, London.

Al-Yom al-Sabi', Paris.

Al-Ahram, Cairo.

Journal of Defence & Diplomacy (Washington),

Aviation Week & Space Technology, London.

Defense & Foreign Affairs,

Flight International, Sutton, UK.

Shu'un Filistiniyyah, Beirut and Nicosia.

Al-Anba', Cairo.

Le monde diplomatique, Paris.

Al-Watan al-'Arabi, Paris.

Interavia, Geneva.

Al-Akhbar, Cairo.

Financial Times, London.

Al-Jumhuriyyah, Cairo.

Middle East Economic Digest, London.

Al-Qabas, Kuwait.

al-Musawwar, Cairo.

al-Mustaqbal al-Arabi, Beirut.

Air International, Stamford, UK.

Middle East Research and Information Project, Washington.

Jane's Defence Review, London.

Jane's Defence Weekly, London.
International Defence Review, Geneva.
Military Technology, Farnborough, UK.
Ar-Ra'i al-'Am, Kuwait.
Al-Ittihad, Dubai.
Al-Qods al-'Arabi, London.
Al-Hawadeth, London.
NATO's Sixteen Nations, Farnborough, UK.
African Defense Journal, Paris.
Survival, London.
Middle East Times, Cairo.
Oman, Muscat.
Naval Forces, Washington.
The Egyptian Gazette, Cairo.
Middle East International, London.
Middle East, London.
International Herald Tribune, Paris.

Index